# Springer Water

The book series Springer Water comprises a broad portfolio of multi- and interdisciplinary scientific books, aiming at researchers, students, and everyone interested in water-related science. The series includes peer-reviewed monographs, edited volumes, textbooks, and conference proceedings. Its volumes combine all kinds of water-related research areas, such as: the movement, distribution and quality of freshwater; water resources; the quality and pollution of water and its influence on health; the water industry including drinking water, wastewater, and desalination services and technologies; water history; as well as water management and the governmental, political, developmental, and ethical aspects of water.

Marzina Begum

# Understanding River Bank Erosion in Bangladesh

## Vulnerability and Coping with Displacement

Springer

Marzina Begum
Department of Public Administration
University of Rajshahi
Rajshahi, Bangladesh

ISSN 2364-6934 ISSN 2364-8198 (electronic)
Springer Water
ISBN 978-3-032-00287-7 ISBN 978-3-032-00288-4 (eBook)
https://doi.org/10.1007/978-3-032-00288-4



This Springer imprint is published by the registered company Springer Nature Switzerland AG
The registered company address is: Gewerbestrasse 11, 6330 Cham, Switzerland

# Preface

Bangladesh, located in the delta of the Ganges, Brahmaputra, and Meghna rivers, is known for its unique landscapes and strong agricultural sector. Nonetheless, the country's dependence on its river systems presents considerable issues, with riverbank erosion becoming a critical environmental issue. Bangladesh's geographical position, characterized by low-lying floodplains and a convergence of major rivers, makes the country extremely vulnerable to flooding and riverbank erosion. In the monsoon season, rivers expand due to the inflow of upstream water, putting significant pressure on the riverbanks and resulting in extensive erosion. This erosion endangers the livelihoods of the riverside residents and significantly jeopardizes the country's agricultural productivity.

Riverbank erosion is a critical environmental issue that has received increased attention in recent years in Bangladesh. Riverbank erosion can result in the loss of agricultural land, damage to infrastructure, and consequently jeopardize livelihoods. Nonetheless, implementing appropriate strategies for riverbank erosion and developing an effective action plan to control and regulate erosion is essential. Riverbank erosion is attributed to various contextual factors, including fluvial erosion and excessive water flow in the river channel. The underlying causes in each occurrence may be either natural or human, or both, and they may alter through time and space. Consequently, there is an increasing acknowledgment of the need to reconcile bank erosion control with the protection of natural riverine ecosystems.

Bangladesh is a country characterized by its many rivers, and the effects of riverbank erosion forced millions to leave their ancestral homes. Annually, riverbank erosion in Bangladesh causes significant death, damage to property and loss of agricultural land. Riverbank erosion in Bangladesh has significant socio-economic consequences that perpetuate national poverty. This book seeks to address questions regarding the patterns of risks and vulnerabilities in areas susceptible to riverbank erosion, analyze the impact of riverbank erosion on various forms of livelihood vulnerability, and reveal the government's challenges in managing the risks and vulnerabilities associated with riverbank erosion.

Rajshahi, Bangladesh
November 2024

Marzina Begum

**Competing Interests** The author declares the following as potential competing interests: NA

# About This Book

Every year, an increasing recurrence of natural disasters, including cyclones, floods, and riverbank erosion, displaces millions of individuals from their houses. Riverbank erosion is a significant geomorphological process that impacts changes in river paths in alluvial plains. It has lasting effects and cannot be naturally reinstated. Bangladesh is a country characterized by its many rivers, and the impacts of riverbank erosion compels millions of people to evacuate their ancestral homes. Every year, riverbank erosion in Bangladesh results in significant fatalities, property damage, and loss of croplands. Riverbank erosion in Bangladesh has severe socio-economic consequences contributing to national poverty. This book aims to find the answers to the following inquiries. (i) What are the key patterns of risks and vulnerabilities in areas prone to riverbank erosion? (ii) What are the impacts of riverbank erosion on different types of livelihood vulnerability? (iii) What are the reasons of the government's failure to deal with the risks and vulnerabilities caused by riverbank erosion?

This book consists of six chapters. Chapter 1 provides an introductory discussion on the patterns and livelihood risks and vulnerabilities of Internally Displaced Persons (IDPs), along with the research methodology to execute the study. It also includes a discussion on the study location, specifically the Sirajganj District, which is greatly affected by riverbank erosion. Chapter 2 extensively examines the theoretical foundation of conducting this research endeavor. This chapter provides a full exploration of the research methodology, such as sample size, data collection instruments, and the approach to conducting the research. Chapter 3 presents theoretical and empirical evidence showing that affected people in the study areas faced multiple factors that made them vulnerable. Therefore, this study proposes the implementation of measures to enhance economic prospects and social advantages for the households of migrants, with the aim of effectively addressing future instances of involuntary migration in Bangladesh. Furthermore, organizations that provide microcredit excluded households without land in the study areas, therefore restricting their capacity to encounter with financial difficulties during periods of financial difficulties. Insufficient access to basic health services and the unwillingness of its professionals to stay in rural places not only lead to negative health consequences but also contribute to life-threatening situations.

The findings of Chapter 4 present that affected people are inclined to stay in the same place and are reluctant to move unless there are major negative circumstances. Despite of the risks and vulnerabilities due to riverbank erosion and the destruction of ancestral home and infrastructure, the residents of riverine regions typically choose to stay in their original location rather than relocating to nearby or distant areas. Occasionally, adult members of households for earning wages travel to different regions of the country, but usually just for short periods of time. These coping and adaptation strategies strengthen the remaining family members to continue living in their permanent places. Further, the comprehensive analysis of the underlying factors contributing to risk and vulnerability reveals that environmental elements are associated with riverbank erosion, coexist with other factors, such as economic, institutional, and infrastructural and collectively have a greater impact on people's livelihood.

Chapter 5 proposes measures that can be beneficial for the development and implementation of policies that are specific to the research objectives. These measures particularly focus on improving the well-being of involuntary migration, and non-migrant who reside along the river in the study areas and other similar locations that are prone to riverbank erosion in Bangladesh. Prior to forced migration, marginalized individuals typically depend on indigenous capacities and strategies for dealing with erosion challenges, the effectiveness of which often relies on government and family support. Although a significant number of these individuals prefer to migrate temporarily, particularly during difficult times, the relatives of these migrants continue to reside in their original locations, whether it be their ancestral homes or newly established residences.

Chapter 6 emphasizes the importance of analyzing the main results of the study and addressing the deficiencies in the existing support programs for the households who have been impacted by river bank erosion in Bangladesh. The findings show that the displaced population due to erosion along the river receives no significant support in supporting livelihood, especially in reconstructing homes and opportunities in income generation.

## Acronyms

1 Sotok = 435.6 sq ft

Bigha is a traditional land measurement unit used across Bangladesh, and land sales are still conducted using this unit. 1 Bigha = 14,400 sq ft (1337.8 sq. m or 1600 sq yd).

District—It is the second administrative tier of local government in Bangladesh.

Jotedar—It means who has more land capital and the ability to borrow some farmers.

Khas land—The owner of land property is not any citizen or private organization and it is under the territory of the nation under Government authorization.

NIRAAD—NIRAPAD, an alliance of 22 non-governmental organizations, aims to mitigate disaster risk in Bangladesh by addressing the lack of information and understanding about disasters and enhancing the disaster risk management capabilities of development organizations and communities.

Upazila—It is the second administrative tier of local government in Bangladesh.

Union Parishad (council)—This is the Bangladesh's smallest rural local government unit.

# Contents

# About the Author

**Marzina Begum** Professor of Public Administration, University of Rajshahi, Bangladesh.

Specialization: Disaster Management, Environmental Policy and Governance, Climate **Change.**

Dr. Mst. Marzina Begum is a Professor in the Department of Public Administration at the University of Rajshahi, Bangladesh. She holds a Bachelor of Social Science (BSS) and a Master of Social Science (MSS) from the University of Rajshahi, Bangladesh. She continued her academic journey by obtaining her Doctor of Philosophy (Ph.D.) from the University of Pavia, Italy. Dr. Marzina Begum's research interests are centered on disaster risk management, environmental policy and governance, and climate change. Her scholarly contributions are substantial, with research articles published in distinguished international journals and chapters in books from renowned publishers such as Springer, Routledge, and Business Expert Press. Throughout her career, Dr. Marzina Begum has actively participated in international seminars, workshops, and conferences, where she has presented her research findings and engaged in discussions on critical issues in her field. She has also been involved in various research projects, demonstrating her dedication to advancing knowledge and addressing societal challenges related to environmental sustainability and disaster resilience. In recognition of her outstanding contributions, Dr. Marzina Begum was honored with the GroBrundtland Award in **2016** by the Tang Prize Foundation in **Taiwan.**

# Abbreviations

| | |
|---|---|
| BBS | Bangladesh Bureau of Statistics |
| BCCSAP | Bangladesh Climate Change Strategy and Action Plan |
| BWDB | Bangladesh Water Development Board |
| CDMP | Comprehensive Disaster Management Programme |
| CDSP | Char Development and Settlement Project |
| CEGIS | Center for Environmental and Geographic Information Services |
| CIID | Climate Induced Internal Displacement |
| CRO | Chief Resettlement Officer |
| DAE | Department of Agriculture Extension |
| DCIID | Disaster and Climate-Induced Internal Displacement |
| DFID | Department for International Development |
| DMF | Displacement Management Framework |
| DPHE | Department of Public Health and Engineering |
| DRI | Disaster Reduction Institute |
| DRR | Disaster Risk Reduction |
| EDPs | Environmentally Displaced Persons |
| FGD | Focus Group Discussion |
| GBM | Ganges, Brahmaputra, and Meghna |
| GR | General Relief |
| HHs | Households |
| IDPs | Internally Displaced Persons |
| IFRC | International Federation of Red Cross and Red Crescent Societies' |
| ILRP | Income and Livelihood Restoration Program |
| IMDMCC | Inter Ministerial Disaster Management Coordination Committee |
| IOM | International Organization for Migration |
| IPCC | Intergovernmental Panel on Climate Change |
| ISDR | International Strategy for Disaster Reduction |
| KIIs | Key Informant Interviews |
| LGED | Local Government and Engineering Department |
| MoDMR | Ministry of Disaster Management and Relief |
| MOEFCC | Ministry of Environment, Forest and Climate Change |

| | |
|---|---|
| MOL | Ministry of Land |
| MVCs | Most Vulnerable Countries |
| MWR | Ministry of Water Resources |
| NDMAC | National Disaster Management Advisory Committee |
| NDMC | National Disaster Management Council |
| NGOs | Non-Government Organizations |
| NSMDCIID | National Strategy on the Management of Disaster and Climate Induced Internal Displacement |
| PD | Project Director |
| PMO | Project Management Office |
| RBA | Rights-Based Approach |
| RP | Resettlement Program |
| RU | Resettlement Unit |
| SDGs | Sustainable Development Goals |
| SLA | Sustainable Livelihood Approach |
| SOD | Standing Orders on Disaster |
| SPS | Safeguard Policy Statement |
| UNEP | UN Environmental Program |
| UNHCR | United Nations High Commission for Refugees |
| UWICED | University of the West Indies Centre for Environment and Development |
| VGD | Vulnerable Group Development |
| WB | World Bank |

# Chapter 1
# Riverbank Erosion in Bangladesh: Research Context and Nature of Inquiry

**Abstract** Riverbank erosion has a negative impact on many countries worldwide, resulting in the loss of land, infrastructure, food supplies, and biodiversity, necessitating the displacement of populations. Riverbank erosion is a frequent occurrence in low-lying deltaic countries, including Bangladesh. River bank erosion is the most detrimental natural disaster, causing significant damage to both the population and their assets. Being a highly erosion prone area, riverbank erosion has disastrous effects on people's livelihood over the time in Sirajganj district which is located along the bank of Jamuna River. However, millions of people depend on the Jamuna river system in Bangladesh for their everyday needs. This chapter focuses on analyzing the patterns and risks and vulnerabilities of people's livelihood due to riverbank erosion in Bangladesh. This chapter will also provide a justification for the research methodology used in this book.

## 1.1 Introduction

Riverbank erosion is a common environmental concern that impacts regions worldwide, having substantial consequences for both natural ecosystems and communities. Therefore, understanding the measures and developing appropriate mitigation strategies is essential for sustainable river management. A primary cause of riverbank erosion is the characteristics of river flows, which may put significant pressure on the banks, resulting in a gradual loss of sediment and eventual breakdown of the bank structure (Florsheim et al. 2008). In many developing countries, the issue is intensified by an interplay of natural and human-induced factors, including increased sedimentation from upstream activities, changes in land use, and the construction of hydraulic structures that impede the normal flow of rivers (Okeke and Ede 2019).



M. Begum, *Understanding River Bank Erosion in Bangladesh*, Springer Water,
https://doi.org/10.1007/978-3-032-00288-4_1

Previous research has examined the intricate interactions among various factors that lead to riverbank erosion, including soil characteristics and the frequency of extreme weather events. A research on the Nkisi River in Southeast Nigeria showed that the riverbanks were unstable, exhibiting a significant potential for sediment loading due to soil characteristics and mass failures (Okeke and Ede 2019). A study on the Sacramento River in the United States emphasized the significance of understanding the synergistic impacts of fluvial erosion and geotechnical instability in influencing bank erosion processes (Ercan and Younis 2007).

To address the issues of riverbank erosion, researchers have investigated several mitigation measures, including the construction of spur dikes that direct and contain the flow while maintaining bank stability (Patel and Kumar 2023). Nonetheless, it is imperative to recognize that not all bank erosion is necessarily harmful, as it can significantly contribute to the ecological integrity of riparian ecosystems. A balanced approach that recognizes the biological importance of bank erosion processes is essential for effective river management (Patel and Kumar 2023).

Bangladesh is a country that has many rivers, with over 220 of them flowing through the country (Rahman and Islam 2016). Out of these, 57 rivers are international, meaning they flow through other countries as well. The total length of the river banks in Bangladesh is 2400 kilometers (km) (Rasheed 2008). Bangladesh is located in the delta formed by the convergence of the Ganges, Brahmaputra, and Meghna (GBM) rivers. It is among the most widespread river systems in South Asia. The Ganges river system has a catchment area spanning 1.09 million square kilometers and it begins at the Gangotri glacier in the Himalayas (Rasheed 2008). The river travels a distance of 2526 km before connecting the Meghna River. It passes through China, Nepal, India, and Bangladesh that make an excellent illustration of an international river (Bhuiyan et al. 2017). Situated in the Ganges-Jamuna-Meghna Delta, this country is well recognized as one of the most disaster-prone regions in the world (Rana and Nessa 2017). Further, this country is very susceptible to floods and riverbank erosion due to its three major river systems, the Padma, Brahmaputra-Jamuna, and Meghna-and their enormous floodplains. Erosion in these areas, however, is a catastrophic occurrence that attracts significant attention in the news.

The issue of river erosion is a frequent occurrence, posing risks and vulnerabilities to 15–20 million individuals and directly impacting one million people in 94 Upazilas per year in Bangladesh. River bank erosion is the most catastrophic natural calamity in the country in terms of the amount of land lost, destruction to property, and displacement of people (Bhuiyan et al. 2017). These devastating environmental events not only result in the internally Displaced Persons (IDPs) within the country, but also lead to sufferings of human life and put economic pressure on the nation (Muinul 1992; Alexander 2002).

Brown (2008) asserts that a clear and precise definition of IDPs is necessary for building a stronger sense of national and global accountability and to highlight the importance of effectively addressing environmental threats. In order to formulate an effective migration policy, it is crucial for policymakers and the affected community to possess an extensive understanding of the interplay between the local environment and internally displaced persons (IDPs). There is a considerable number of

IDPs in Bangladesh who are compelled to abandon their ancestral homes owing to river bank erosion. The hazard primarily affects places with flood lines and steep slopes, which pose a significant threat to people's livelihoods. Additionally, it is important to acknowledge that forced migration or displacement can be induced by factors other than just environmental degradation. According to Terminski (2012) and McGregor, IDPs are people who are forced to move or abandon their homes that pose a threat to their way of life, whether these changes are temporary or permanent. These changes entail increased volatility in seasonal weather patterns, such as the monsoon. Consequently, individuals may seek refuge in different regions, especially in their capital city. However, in Bangladesh, IDPs have emerged as a significant issue of national importance.

This research operationalizes vulnerability as the degree to which the environment is impacted by degradation caused by external influences (UWICED 2002). Vulnerability to climate change, as operationalized by the IPCC, refers to the degree to which a system is prone to or incapable to effectively handle the adversarial effects of resultant changing climatic conditions. It also refers to the character, extent, and variability of climate change that a system is vulnerable to, as well as the effects on people's livelihoods and their ability to adapt (IPCC 2007). Nevertheless, the level of vulnerability differs depending on the location and period of time, as well as the policy and its actions inside the context of a particular state. The research findings and policy materials emphasize the vulnerability and the perceived and actual threats to livelihood. Undoubtedly, many risks and vulnerabilities arising from climate change, as seen by one community, may be of little importance to another due to factors of context (IPCC 2007). Considering the matter, it is contended that one of the concepts relating to climate change is the imperative of survival, which can vary the level of urgency depending on the place, time, and intensity of the dangers. Accordingly, Brauch et al. (2009) rightly states that there is a lack of agreement on the definition, criteria, and indicators used for assessing vulnerability (Brauch et al. 2009).

Only in June 2021, river bank erosion caused a significant number of internally displaced persons (IDPs) in five districts: Sirajganj, Jhalokathi, Kurigram, Gopalganj, and Patuakhali. This resulted in the destruction of 315 houses and 2 schools, as well as severe damage to 13 villages only in the Sirajganj district. The communication network between 8 villages in Dashmina upazila of Patuakhali district was disrupted as a result of the damage to a 1.5 km road (NIRAPAD 2021). Due to its great susceptibility to erosion, riverbank erosion in Sirajganj district, located along the Padma River, has devastating consequences on the local population's livelihoods over time. A vast number of individuals rely on the Padma River System in Bangladesh to fulfill their daily need. This research focuses on analyzing the patterns and vulnerabilities of river erosion on people's livelihoods. This study aims to determine the answers to the following inquiries. (i) What are the patterns of risks and vulnerabilities in riverbank erosion prone areas? (ii) What are the effects of riverbank erosion on different forms of livelihood vulnerability? (iii) What are the reasons for the government's inability to effectively address the vulnerabilities caused by riverbank erosion?

## 1.2 Research Context

Norman Myers, a renowned British researcher in the domain of environment and climate change, authored a report on the phenomenon of environmental exodus in response to the current global crisis. While Myers has previously written extensively on environmental change and migration caused by displacement, his report has come under criticism from Vikram Kolmannskog for its lack of consistency and failure to address strategies to encounter with climate change (Myers and Kent 1995). According to Myers and Kent (1995), it is projected that by 2050, China would have around 30 million, India 30 million, and Bangladesh 15 million IDPs respectively. The rise in the number of IDPs occurs in various locations worldwide due to multiple factors, such as desertification, water scarcity, salinization of water and land, and the damage of biodiversity.

Richard Black conducted comprehensive research on IDPs in Africa and other regions. In his research for the UNHCR titled "Environmental refugees: myth or reality?", Black challenges the concept of environmental refugees by asserting that while the environment does contribute to forced migration, there are always additional factors at play, including political and economic influences (Black 2001). He strongly dismisses the notion of climate refugees, deeming it an illusion as well as a deceptive concept, which has now become heavily politicized among world leaders (Black 2001). Black disagrees with Myers' beliefs and conclusions, and even characterizes them as a neo-Malthusian approach that relies on questionable assumptions (Black 2001). He asserts that these displaced people pose a direct threat to local security. He asserts that there is insufficient evidence to specifically attribute environmental changes as a direct cause of large-scale forced migration to wealthy countries.

Bangladesh is located within the basin of the Ganges, Brahmaputra, and Meghna (GBM) rivers. The GBM basin has had the greatest deposition of sediments on record. It is estimated that the Bengal basin releases around 1050 million tons of silt each year, with 600 million tons being deposited in the Bengal delta. In the end, riverbeds in GBM become filled with silt and lose their depth. The river morphology undergoes frequent changes, with the river channel shifting periodically due to excessive silt accumulation (Bhuiyan et al. 2017).

Bangladesh is classified as one of the MVCs when it comes to climate-induced disasters. From 1998 to 2017, Bangladesh experienced several extreme weather occurrences, ranking it eighth among the countries most impacted, according to Eckstein et al. (2019). Bangladesh has been identified and classified by the IPCC and the WB as one of the MVCs to flooding, cyclones, and other related disasters (Mehedi et al. 2010). The monsoon period in Bangladesh is featured by several types of natural disasters, including riverbank erosion, which are influenced by the riverine geography of the country (Etzold et al. 2014). Such events are shown the occurrence of the country's fluvial flooding and riverbank erosion. Data on erosion and deposition along the Ganges river between 1973 and 2011 indicates that the left bank gained around 59 $km^2$ of land, while the right bank had a loss of approximately

57 km$^2$ (Dewan et al. 2017). Following a dangerous event, which often exceeds the ability of households living in riverine areas to cope and get government assistance, it becomes difficult for them to sustain their livelihoods, eventually leading to the decision to migrate. According to Gravgaard and Wheeler, riverbank erosion is the sole underlying causes of displacement in Bangladesh, with an annual range of fifty to two hundred thousand (50–200,000) individuals.

BCCSAP stated issues and concerns about forced migration and highlighted efforts to take appropriate action to safeguard people's livelihoods (Government of Bangladesh 2009; Mehedi et al. 2010; Rabbi et al. 2013). Christian aid, a British support charitable organizations, projects that by 2050, some 250 million individuals worldwide may experience displacement due to direct consequence of climate change, necessitating migration from their traditional habitat (Kabir and Kamruzzaman 2022). According to Myers, the estimated number of internally displaced persons (IDPs) is approximately one in every forty five individuals worldwide. However, Bangladesh is considered to be unable in addressing the needs and implementing measures for IDPs due to constraints in government expenditures and its high population density. In addition, there has been less academic attention paid to the complex characteristics of vulnerability and how it affects marginalized internal migrants in riverine areas (Elahi and Rogge 1991).

The inhabitants along the river bank areas are considered to be the most marginalized demographic in society and are not sufficiently paid to attention by the scientific community. Prior studies on riverbank erosion have examined its status in simple from a geographical and geological perspective, however, this book investigates the most vulnerable riverbank erosion area along the Padma River near Chauhali and Shahajadpur upazila, and its effects on the socioeconomic status of the local population.

Another challenge that must be tackled is the support of the government and administrative capacity. To achieve this, it is necessary to allocate more public funds towards building civic infrastructure and managing critical services, such as housing, water, and other basic facilities. This is particularly challenging in a poverty-stricken country like Bangladesh, where ensuring the well-being of its citizens is a struggle. Recently, the topic of IDPs has attracted national attention from policy-makers, environmental advocates, and social scientists. Consequently, it necessitates careful consideration, a comprehensive understanding, and an awareness of the potential societal consequences (Momen 2021). This book aims to investigate the potential relationship between riverbank erosion and IDPs in the most vulnerable river bank erosion area in Bangladesh, with the intention of supporting future disaster management strategies.

This book investigates the current status of IDPs in Shahjadpur Upazila and Chauhali Upazila, located under the Sirajganj District, as the most river bank erosion prone areas in Bangladesh. As one of the MVCs, Bangladesh must do research to implement improved river bank management strategies. These strategies can then be integrated into national policies to produce rules for the resettlement of IDPs. This book examines the factors that lead to IDPs by investigating the consequences of riverbank erosion where these individuals seek shelter after leaving their homes of origin. Furthermore, this book can provide valuable insights to environmental activists and policy makers regarding socio-economic conditions and the specific initiatives required to guarantee fundamental livelihood rights and effective management of key state services for IDPs. Notably, there is no significant comprehensive research on displaced people and their livelihood risks and vulnerabilities in Bangladesh due to riverbank erosion. Specifically, there is limited information on how affected individuals perceive their own vulnerability due to river bank erosion, despite the likelihood of more displacement events occurring in the future (Kabir and Kamruzzaman 2022). The response provided by the government in addressing riverbank erosion is inadequate (Dewan et al. 2017). However, the term "environmentally displaced persons" is often used to refer to climate change-induced forced migrants, as recognized by the International Organization for Migration (IOM) and other worldwide organizations. This term aligns with the objectives of the UNHCR's IDPs. Hence, it is imperative for the government of Bangladesh to give utmost importance to addressing the challenges experienced by individuals who are forced to abandon their ancestral homes due to the detrimental consequences of river bank erosion.

Just in a recent incident, for 10,000 bags of sand were dumped to prevent the erosion along the Jamuna River in Sirajganj. Out of the 11 family members of the dead, two individuals were given a sum of BDT 20000 (Available at https://modmr.gov.bd July 1, 2021). However, the primary aim of the book is to identify the patterns of livelihood risks and vulnerabilities in Shahjadpur and Chauhali Upazila, located under the Sirajganj District as the most vulnerable areas of the Bangladesh due to river bank erosion. These individuals were compelled to leave their ancestral homes due to riverbank erosion. The book aims to analyze the immediate and long-term societal impacts of displacement, and assess the administrative capacity and government support in dealing with this disaster.

## 1.3 The Existing Literature on River Bank Erosion: Causes, Impacts and Measures

Riverbank erosion is a widespread environmental issue impacting livelihood globally (Walsham 2010). This phenomena concerns as it may result in the loss of valuable land, damage to infrastructure, and disruption of local livelihoods and ecosystems. Different factors and issues contribute to riverbank erosion, comprising both natural and human-induced influences. The earlier research indicates that in the Asia and Africa, the transformation of forests into agricultural land have intensified surface runoff, resulting in increased riverbank erosion and related climate risks, such as flooding and surface water contamination. High-intensity rainfall, alterations in water flow, and the intrinsic soil characteristics of the banks might intensify the erosion process. Furthermore, anthropogenic activities like alterations in land use, groundwater extraction, and the degradation of riparian vegetation have been associated with increased riverbank erosion in many areas (Okeke et al. 2020; Chuenchum et al. 2019; Nde et al. 2022).

Riverbank erosion is a complicated issue that necessitates a global perspective and coordinated efforts for mitigation measures. By understanding the principal causes of this problem and executing various mitigation techniques, it is feasible to reduce the effects of riverbank erosion and foster the sustainable management of river systems globally. However, it is understood the need of a comprehensive strategy for mitigating riverbank erosion, stressing the necessity for cooperation among many stakeholders, including governmental bodies, local communities, and environmental organizations. By tackling the root causes of riverbank erosion and executing a thorough range of mitigation strategies, long-term sustainability of river systems and the people reliant on them can be promoted.

A comprehensive strategy is necessary to tackle the issues of riverbank erosion, incorporating both structural and non-structural interventions. Structural interventions may encompass the construction of riverbank reinforcements, like retaining walls, to stabilize the banks and mitigate the effects of water flow. Non-structural measures involve the adoption of sustainable land use practices, the vegetation along rivers, and the advancement of integrated river bank management systems as well (Okeke et al. 2020).

Riverbank erosion constitutes a critical environmental challenge in Bangladesh, especially in the major river systems of the Ganges, Brahmaputra, and Meghna basins. The alteration and redirection of river flows significantly contribute to bank erosion. The high sediment load in the rivers, as the deposited sediments can modify river morphology and flow patterns, resulting in increased erosion (Bhuiyan et al. 2017). Alongside natural factors, activities caused or influenced by people have also expedited riverbank erosion in Bangladesh. Unsustainable extraction of sand from riverbeds, an improper strategies for building and filling materials, can disturb natural river dynamics and exacerbate erosion (Islam et al. 2018).

This continuing process is primarily attributed to the alteration of river routes, with some contribution from flooding. Though Hossain (1993) argued that the impact of riverbank erosion has not yet been properly acknowledged, both in Bangladesh and beyond. The significance of the issue is apparent from estimates indicating that around 1050 million tons of sediment are discharged annually from the Bengal basin, resulting in substantial losses of property, infrastructure, and agricultural land along the riverbanks. This study seeks to examine the potential causes of riverbank erosion in Bangladesh and evaluate the strategies that might be implemented to minimize its effects (Hossain 1993).

Riverbank erosion has enormous economic and social effects on the communities residing along the riverbanks in Bangladesh. Erosion results in the degradation of agricultural land, residences, infrastructure, and other significant assets, leading to economic distress and displacement for the impacted households. The reduction of agricultural land may lead to diminished productivity of crops, hence intensifying the financial burden on the local population.

Besides the economic effects, riverbank erosion can also result in social and environmental consequences. The loss of land and the displacement of populations can promote tensions and conflicts, as individuals endeavor to secure new residences and means of livelihood.

Riverbank erosion is a multifaceted and varying problem in Bangladesh, resulting in substantial economic, social, and environmental repercussions. Resolving this difficulty necessitates a holistic approach that accounts for both natural and human-induced elements contributing to the issue. Through the implementation of both structural and non-structural interventions, Bangladesh can mitigate the effects of riverbank erosion and enhance the long-term resilience of its riverine communities.

A primary structural measure is the construction of embankments and other kinds of structures to stabilize riverbanks and mitigate erosion. Dredging and sediment management techniques, including the strategic allocation of dredged materials, can assist in preserving river shape and mitigating erosion. Alongside structural solutions, non-structural strategies, including the adoption of sustainable land-use practices, reforestation, and the establishment of early warning systems, are vital in minimizing the effects of riverbank erosion.

Riverbank erosion presents a major environmental issue in Bangladesh, especially in the Sirajganj region, where the Jamuna, Padma, and Meghna rivers interconnect. This process, influenced by numerous natural and human-induced factors, consistently endangers the lives and infrastructure of the local population. The sediment-rich rivers of the Bengal delta, including Sirajganj, are among the most dynamic globally, undergoing swift alterations in their courses and bank locations. Bhuiyan et al. (2017) in his research understanding the reasons and formulating efficient strategies for decreasing riverbank erosion is essential for sustainable development and safeguarding at-risk communities in the area.

Riverbank erosion in Sirajganj is a multifaceted process shaped by various sources. The substantial sediment load transported by the rivers is a significant factor in this process, with the Ganges, Brahmaputra, and Meghna basins projected to release over 1050 million tons of silt per year (Bhuiyan et al. 2017). The alteration and diversion of river flows, intensified by floods, further exacerbate riverbank erosion (Hossain 1993). Moreover, human activities like sand dredging from riverbeds have been shown to affect the stability of riverbanks, resulting in increased erosion in some regions (Islam et al. 2018).

Riverbank erosion in Sirajganj, Bangladesh, presents a difficult challenge necessitating a thorough understanding of its underlying causes and the execution of suitable mitigation strategies. By tackling the causes of erosion, both natural and human-induced, and implementing a comprehensive strategy for river management, the detrimental effects on local residents and infrastructure can be diminished, facilitating sustainable development in the area (Hossain 1993; Bhuiyan et al. 2017; Okeke and Ede 2019; Islam et al. 2018).

Most climate-related displacement and migration occur within national boundaries, since 2008, an annual average of over 20 million people have been internally displaced annually by weather-related extreme events, with storms and floods being the most common (IPCC 2022). Mutton and Haque and Stern predicted that China would experience the displacement of 30 million individuals, India 30 million, Bangladesh 15 million, Egypt 14 million, other delta areas and coastal zones ten million, island states one million, resulting in a total of 50 million displaced people by 2050.

Myers note that the concept of 'climate refugee' gained significant media and scholarly interest due to the forecast that by 2050, the global population of displaced individuals might exceed 200 million. In another research, this includes an estimated 73 million internally displaced persons in China, 26 million in Bangladesh, and 20 million in India (Kabir and Kamruzzaman 2022). Rahman examines the socio-demographic characteristics of individuals affected by riverbank erosion in the Jamuna floodplain areas in Bangladesh, in his research article titled 'Impact of Riverbank Erosion Hazard in the Jamuna Flood plain Areas in Bangladesh'. A significant number of individuals are compelled to evacuate their residences and find shelter on roadside embankments, adjacent properties, and ancestral property due to riverbank erosion.

Islam et al. conducted a study named "Bank Erosion of the River Meghna: Population Displacement and Socio-Economic Impacts," in which they observed that Bangladesh experienced the adverse effects of Meghna river bank erosion, resulting in forced displacement and socioeconomic consequences. Based on the facts, riverbank erosion significantly impacts the population, leading to the complete loss of their property. Thus, there were socio-economic difficulties and forced relocation that impacted individuals and households. This had consequences for various aspects like the size of households, educational achievements, and involvement in the workforce, occupational positions, property ownership, and income. Multiple studies found a potential association between the occurrence in river erosion and the increased number of urban population.

After the emergence of Bangladesh as an independent country, Elahi, Islam, and Khan conducted studies which indicated that the rate and pattern of urbanization has experienced a growth of over 60% per decade. This increase can be attributed mostly to the migration of people from marginalized rural areas to metropolitan city, rather than a regular growth in the urban population. Hutton and Haque discovered that urbanization has expedited the proliferation of colonies and squatters in several urban regions. According to Akhter (1984), around 25% of slum residents relocated from rural regions to urban areas, with a particular focus on Dhaka city, as a result of natural calamities. Hossain (1984) and Uddin and Basak (2012) calculate that 10% of riverbank erosion led to the migration of marginalized individuals to metropolitan regions in pursuit of a viable means of supporting themselves.

One decade back, riverbank erosion impacted 94 out of 489 Upazilas (subdistricts), leading to IDPs and increased risks and vulnerabilities among inhabitants living along the rivers in Bangladesh. While riverbank erosion is a matter of concern, there have been limited academic studies that have explored the intricate ways in which vulnerability impacts the lives of the underprivileged and marginalized, as well as their non-migrant households who continue to reside in locations prone to erosion (Uddin and Basak 2012). According to Adger et al. (2015) and Kabir and

Kamruzzaman (2022), it is scientifically believed that climate change and natural disasters have substantial impacts on human vulnerability, which in turn affects people's migration or forced displacement.

In his study titled, Impact of River Bank Erosion on Socioeconomic Condition: An Examination of Displaced Population in Sadar Upazila, Gaibandha District, Bangladesh, Nirmal sought to demonstrate the socioeconomic effects of erosion in the Sadar Upazila of Gaibandha District. Bangladesh possesses an extensive system of rivers and canals, with the majority being ephemeral in nature. Consequently, the erosion of river banks is a frequent occurrence. Annually, river erosion, soil degradation, and flooding displace three million individuals. Therefore, the author was needed to look into the nexus between river bank erosion and its effects on the livelihood vulnerabilities of the affected individuals in Bangladesh who have been forced to relocate in the new areas.

Baki Abdullahel (2014) conducted a study named 'Socio-Economic Impacts of Gorai Riverbank Erosion on People: A Case Study of Kumarkhali, Kushtia', which involved both qualitative and quantitative research approaches. According to the findings, individuals who have been affected have experienced substantial socioeconomic disadvantage. The unfavorable impact affects 93% of individuals, whereas only 2% experience a good impact. The population of Kumarkhali have had substantial socioeconomic consequences due to the erosion of the Gorai riverbed, which continues to have a detrimental influence on their means of living.

Nevertheless, prominent scholars have proposed that internal displacement driven by some underlying reasons have evolved beyond traditional "push" theories and for a deeper understanding of socio-environmental disparities (Hunter et al. 2015). Etzold et al. (2014) investigated the impact of changing rainfall patterns in northern Bangladesh and observed that social inequality and poverty have a greater influence on internal migration than bio-physical causes. In his work "Hazards in a fickle environment: Bangladesh," Chowdhury presents a comprehensive and first-hand longitudinal investigation of the risks associated with rivers in Bangladesh. This study examines the attributes and essence of reactions to the dangers of floods and riverbank erosion, together with an assessment of the impacts caused by these hazards. The book states that social, economic, and political factors, along with the unpredictable physical environment of the floodplains, all have a role in the impact of river bank erosion.

The ecological sensitivity and dynamics of the Ganges-Brahmaputra deltaic system, along with the sociocultural adaptations of people to these riverine events, are considered.

Elzbieta and Dianna examined the topic of environmentally forced migrants among the inhabitants along the rivers in Bangladesh and their ability to reconstruct space and place in challenging conditions which was included in the edited book "Reconsidering Refugee and Displacement" by Doreen Indra. This chapter explores several challenges associated with representing Bangladeshi environmental migrants. Several factors have contributed to the discourse around individuals displaced by riverbank erosion, which mostly associates them with periods of crisis, presents them as more uniform than they truly are, and depicts them as passive victims of uncontrollable natural forces (Piguet 2013). The allocation of scarce aid and development funds is largely driven by the symbolic impact of these recurring images of helpless, and powerless forced migrants. She enumerates the "pre-adaptations" that the inhabitants of Kazipur have developed in response to rapid change and devastation caused by river bank erosion. Finally, the author illustrates how affected individuals take independent actions and make strategic decisions when it comes to resettlement. They also explore the wide range of adaptations made in terms of production and social reproduction, although facing limitations.

In their policy brief titled 'Coping with Riverbank Erosion-Induced Displacement', Abrar and Azad (2007) focus on the impacts of riverbank erosion on women and provide strategies to address this unfortunate situation. Irrespective of circumstances, women experience the most hardships in the country. Affected individuals residing in shelter camps encounter significant obstacles. Even in the camps, these vulnerable women are prone to sexual assault. According to the World Bank's estimation, sexual and gender-based violence is responsible for 5% of the overall disease burden among women of reproductive age in developing nations.

In his work titled 'Mental health and the psychosocial consequences of natural disasters in Asia', Udomratn (2008) presents a comprehensive analysis of natural disasters in Asia, along with an examination of mental diseases and psychosocial treatments associated with these disasters. According to this study, the occurrence of post-traumatic stress disorder and its symptoms among Asian survivors of natural disasters varies from 8.6 to 57.3% (Udomratn 2008). Thailand's data shows that children affected in posttraumatic stress disorder (PTSD), as diagnosed by psychiatrists, and the symptoms of PTSD in adults, as measured by sample question, decreased over time. However, the rate of mental depression related disorder in children remained unchanged, while depression in adults presented a slight fall after nine months of follow-up. In the developing nations in Asia, there is a shortage of mental health experts and lack of understanding and implementation of healthcare for affected people due to disasters. This can result in delays in providing psychosocial interventions and rehabilitation for victims (Udomratn 2008; Momen and Ferdous 2023).

According to Patwary (2016), in his study titled "Climate Change, Human Displacement and the Potentiality of Conflict: The Case of Bangladesh," he highlighted that the displacement of people due to natural disasters, whether they occur suddenly or gradually, is a highly apparent outcome of climate change in recent times. Historically, extreme climatic events have caused significant damage and disruption to the livelihoods of millions of people along the rivers throughout the country. Lives become difficult in different disaster prone areas, causing more people to be forced to leave the coastal parts of the country. This situation has also caused significant pressure on the government's limited capacity and has a cause of livelihood risks and vulnerabilities. Within this particular context, the book aims to comprehend the mechanisms through which resultant river bank erosion leads to the displacement of individuals from their ancestral homes.

In their study titled "Riverbank Erosion-Induced Stress and Coping of Displaced Women in Bangladesh," Keya and Harun (2007) investigated the psychological stress experienced by women who were forcibly displaced due to riverbank erosion. The objective of this study was to assess the psychological stress levels and coping strategies employed by women impacted by river erosion, who were forced to relocate to government-owned khas land due to the loss of their homes and property. Displaced women employed various coping strategies, including seeking social support. Women utilized less confrontational and less aggressive coping techniques, which suggests a favorable adaption strategy. Elahi et al. (1991) found that women who had been displaced revealed lower levels of self-control, acceptance of responsibility, and planned problem-solving coping mechanisms than women who had not experienced displacement.

In their publication, Uddin and Basak (2012) examined the effects of riverbank erosion on the livelihoods residing along the riverbanks. The research was carried out in the most disaster prone areas of Bangladesh affected by riverbank erosion, specifically in the Kapasia Union under Gaibandha District and Kazipur and Khasrajbari Union under Sirajganj District. The objective of the study was to investigate the strategies employed by the local inhabitants to deal with riverbank erosion and assess its effects on their livelihood vulnerabilities.

Haque (1987) conducted an analysis of the Lower Brahmaputra (Jamuna) floodplain in, focusing on the overall effects of disasters. An assessment has been conducted to determine the vulnerability for individuals, focusing on the impact of losses of land, occupational changes, and the effects on their kinship, social ties, and their relationships. A random sample of 547 displaced households was surveyed. A considerable number of households, whose main profession was agricultural labor, were moved as a result of river bank erosion. The article found that the Bengal Alluvion and Diluvion Regulation, 1825 was in effect, however, its purpose was not to establish a new legal principle. Instead, it affirmed the applicability of existing rules and customs. As per the legal instrument, any claims and disputes about newly formed char lands were decided considering the local utilization of payasty (alluvion) and sikosty (diluvium) (Haque 1987). In 1950, the Pakistani government passed a law that ended the permanent settlement with landowners. The law focused on recognizing the de-facto tenure of land that was occupied by Muslim peasants. Additionally, it aimed to provide legal rights for impoverished landowners with smaller plots of land. The power structure is still predominantly controlled by the broader landowning class (jotdar), who also have the authority to decide the ownership of land that appears from riverbeds. Nevertheless, it is crucial to acknowledge that the process of newly accreted land reaching its maximum production capacity might span up to 15 years (Haque 1987). On the other hand, land lost to erosion sometimes consists of highly valuable agricultural land.

Meyer (no date) has made a distinction between mobility, which pertains to movement within boundaries, and migration, which occurs when persons beyond these limits. The divisions can be categorized as geographical, structural, racial, or other factors that the actor acknowledges as distinguishing him from the group. Upon relocating to the metropolis, individuals encounter challenges in adapting to the new surroundings and navigating their way around. This mobility may require a fresh lifestyle, novel beliefs, and new patterns of conduct. The majority of definitions of migration primarily emphasize the act of physically relocating from one geographic area to another. However, there is a distinction in the way migrants are perceived depending on whether they are voluntary or forced.

The aforementioned research examined several facets of livelihood vulnerabilities resulting from river bank erosion in the selected areas of Bangladesh. However, none of the research focuses especially on the assessment of the administrative capabilities and support provided by the government to manage the risk of vulnerabilities. From this perspective, it is necessary to conduct deeper investigation into the current research. However, this book has examined the dependent factors and has now selected the independent variables to investigate their impact on the dependent variables. The variables encompass bio-physical/environmental factors, including soil degradation, economic factors, including income, savings, and credit access, and socio-economic stressors. This book examine whether multiple stressors might have a detrimental impact on certain communities or vulnerable households. An analytical framework has been built for the continuation of research work based on these variables for this study using Fig. 1.1.

The objective of the theoretical framework is to operationalize the key concepts of the study, as well as provide a description of the dependent and independent

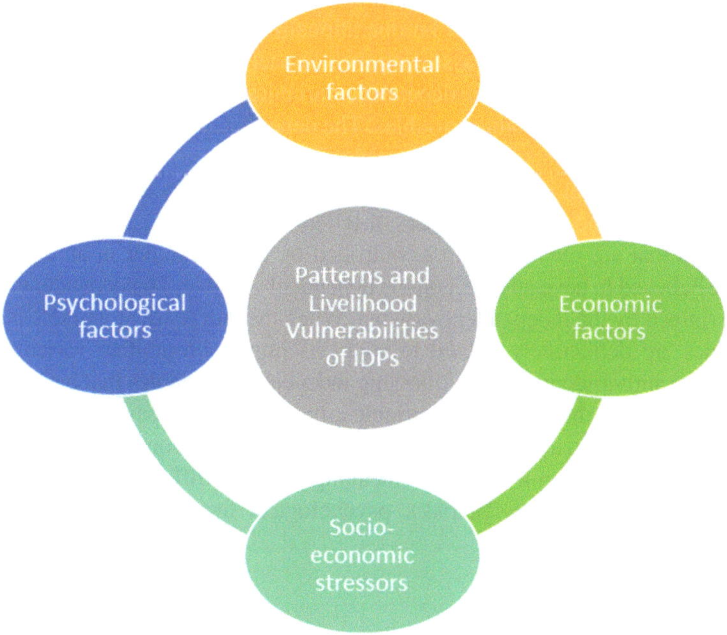


**Fig. 1.1** Interactions between the dependent and independent variables (The author)

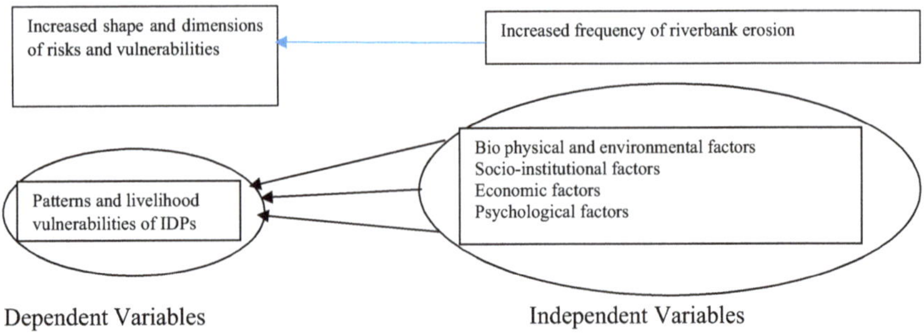


**Fig. 1.2** Theoretical understanding between dependent and independent variables (The author)

variables. A conceptual model has been devised to provide the researcher with a systematic approach to conducting research, based on the discussion. This framework is to define and implement the concepts used in this study, as well as provide a detailed explanation of the dependent and independent variables. A conceptual model, depicted in Fig. 1.2, serves as a guide for researchers to perform their research in a systematic and organized manner. The graphic below illustrates the theoretical nexus between dependent and independent variables.

The term variable is widely used in social science research endeavors. It is crucial to define and identify relationship accurately while developing research objectives (Kaur 2013). Typically, a variable is an attribute whose value can change (Kerlinger 1983). Furthermore, it can be described as a coherent collection of attributes (Earl 1983). Variables can be operationalized by defining them in terms of observable characteristics and assessing their impact on a study, such as distinguishing between independent and dependent variables. However, variables that are influenced or affected by other variables. The dependent variable is the variable that is influenced by the independent variable. The dependent variable is contingent upon the independent variable (Kaur 2013). This research focuses on analyzing the patterns and livelihood vulnerabilities of IDPs.

Factors that are manipulated or controlled by the researcher in an experiment, and are believed to have an effect on the dependent variable. The dependent variable is determined by the independent variable. When the independent variable is active, its values are deliberately altered to study its impact on another variable (Kaur 2013). The author identifies the following independent variables that have a significant influence on the dependent variables: bio-physical or environmental factors (i.e. soil degradation, and shelter); economic factors (i.e. income, savings, or access to credit); socio-economic stressors (meaning that how different factors negatively impact on the community at risk); psychological factors (fear, mental illness, trauma, frustration, etc.).

## 1.4 Methodological Discussion

Qualitative research provides a true interpretation of the research that helps to comprehensively analyze the findings. This research findings approach is mostly descriptive, in nature. Both the qualitative description and quantitative measurements are considered reciprocally in analyzing the social reality of the livelihood risks and vulnerabilities due to river bank erosion that displaced the large number of affected individuals have to experience and to cope with. It is noteworthy that the collected data during the field work are analyzed in descriptive and qualitative manner. However, this study is aimed to establish the analytical and critical examination of both primary and secondary data.

### *1.4.1 The Techniques Followed*

This research followed specific methods to obtain relevant data and fulfill the objectives. This research actually, deal with the vulnerable people and IDPs of Sirajganj district, resultant from riverbank erosion who has taken shelter mostly in other places in Sirajganj district and Dhaka megacity after forced displacement from their ancestral home.

Information gathered from both primary and secondary sources for fact-checking the reliability of data, thus drawing an accurate interpretation. Notably, primary sources of data gathered from semi structured interviews, FGDs and several case stories from the IDPs in the study area. Conducting a case study research, interview served as an important and relevant source of information. Secondary information gathered from published literature, such as books, journals, relevant government websites (Ministry of environment), national environment organizations, BWDB, the concerned Upazila Administration, the knowledge domain of Disaster Management Bureau (DMB) and research report of Ministry of Disaster Management and Relief (MODMR) in Bangladesh and Ministry of Water Resources Management (MOWRM).

However, this research obtained a grant of approximately 2000 US$ dollars from the Faculty of Social Science of the University of Rajshahi for the collection of primary data. This research conducted three-month field study in two years (2022 and 2023) and one month from Dhaka city where the affected households migrated in these two years. By which this research explored the level of vulnerability means, how people were affected by the effects of riverbank erosion and to look at whether it was effectively managed by the concerned body or authority. Data Triangulation method that refers to the use of multiple sources of data in a study is used in this study to enhance validity and to generate in-depth understanding about the issue under investigation (Bryman 2012). In addition this method helped much because, this study had used different qualitative approaches for data collection which generated different perspectives.

### 1.4.2 Target Population

Appropriate selection of the respondents is a significant issue in social science research. Selection of participants for the interview was determined on the relevance of their skills knowledge and direct linkage of the research problem.

### 1.4.3 Data Collection

This study is based on the primary data collection through direct interview selecting purposefully 66 individuals from 54 families, including 20 households from Dhaka urban slum (Dhaka railway station slums who are affected with river bank erosion). In order to better understand internal migration and how they respond to various risks and vulnerabilities due to river-bank erosion. However, the group of persons ranged in age from 20 to 75. The respondents worked as farmers, day laborers, business owners, employees, housewives, and teachers, among other jobs. Apart from interview questions, data collected techniques were used through conducting Focus Group Discussion (FGD) and researcher's personal observation and some case stories of the selected respondents (explorative). However, 2 Upazilas (Shahajadpur and Chouhali) under Sirajgonj district was deliberately selected as study locale due to the most river bank erosion prone areas in Bangladesh. Koizuri and Garadaha are Unions under Shahjadpur upazila, however, Shahjadpur Upazila is considered as one of the river bank erosion prone areas along the Jamuna River. The researcher also collected field data from Gharjan and Omarpur unions in Chauhali Upazila under Sirajganj district.

In-depth interviews were done in different research sites using a predesigned semi-structured questionnaire to explore the reasons of riverbank erosion, losses occurred, ways to cope, and the livelihood pattern. In this research, some of the respondents (12) were from the government officials from the BWDB, the DPHE, Food and Agriculture Department, among others. Further, the researcher interviewed civil society members at the community level, community people and environmental scientists and researcher to cross check the validity and reliability of data. The researcher also talked with representatives from regional non-governmental organization in order to identify the area's most negatively impacted by riverbank erosion.

Key informant interviews (KIIs) (7) are in-depth discussions with professionals, such as community leaders, experts, among others who have firsthand knowledge of the subject under consideration. Key informant interviews are used to gather data on a specific topic, comprehend the nature of problems, and give recommendations for remedies. Seven important informants were questioned in order to have a better understanding of the reasons behind riverbank erosion and the issues that people confront. Primary data were gathered through semi-structured interviews, FGDs, and KIIs. The selection of participants for interviews proceeded with the development of a close relationship between community members and data enumerators. In practice, short-term migrants were located in close proximity to their original village, whereas long-term migrants were predominantly residing distant from their home and currently reside in Dhaka city. Subsequently, a 'snowball' sampling technique was employed to identify the urban population residing in slums. Conducting interviews with the individuals posed a certain level of difficulty due to the significant number of households that had been displaced, but had not resided in the same location for an extended period of time. Therefore, only one adult member of each household was chosen for semi-structured interviews.

The study deliberately chose seven key informants with diverse professional backgrounds, such as one local government official, one representative from a local non-governmental organization (NGO), one Union Parishad (UP) member from each union, and two knowledgeable community leaders from each union. At the time of interviews, respondents were requested to reflect on the primary challenges and relevant issues contributing to their vulnerability that impact on themselves and their family members, as well as the techniques they employ to cope with these issues. Key informants were requested to provide their perspectives on comparable matters, either from their professional expertise or personal encounters with the riverine community.

### *1.4.4 Selection of the Sample Size, Site Visit and Criteria for Selection*

The categories of the respondent were chosen on purpose. Sirajganj district was selected for this study. The study location is the most vulnerable river bank erosion area. Different five categories of respondents were selected purposively. In order to obtaining an overview of research project, some Key informants like, environmental scientists, and academicians were also chosen for, interviewer selected 73 respondents for primary data collection. Among them 12 respondents (government officials) were chosen for an in-depth interview. In addition, 7 key informants were purposively selected to obtain a comprehensive realization of the subject matter of the research. However, during data collection period, the author (in Table 1.1) attempted to hear any opinions and perceptions that might be pertinent to the research and ensure that the study reached the 'data saturation stage'.

Figures 1.3, 1.4, 1.5 and 1.6 shows the trends of riverbank erosion tendency, Sirajganj district and Shahjadpur Upazila and Chauhali Upazila respectively.

**Table 1.1** List of interviewed respondents

| Sl. No. | Patterns of interviews | Number of interviews | Category of respondents | Location |
|---|---|---|---|---|
| 1 | Interviews (from 34 households) | 34 Individuals | Farmers, day laborers, business owners, employees, housewives, and teachers | Sirajganj |
| 2 | Interviews (from 20 households) | 20 Individuals | Rickshaw puller, garments worker, car driver, house servants, paint worker, construction labor | Dhaka |
| 3 | In-depth interviews (semi-structured questionnaire) | 12 Individuals | BWDB, DPHE, Food and Agriculture Department, Department of Social Welfare, Department of Family Planning, Fire Services | Dhaka and Sirajganj |
| 4 | Key Informant Interviews (KIIs) | 7 Professionals | Community leaders and experts | Dhaka and Sirajganj |
| 5 | Case stories (12 households) | Representatives from 12 households (one representative from each household) | Community from the most vulnerable households | Sirajganj |

## 1.5 Relevant Documentation

Both sources of primary and secondary data collection strategies were employed for the study. Qualitative data and its analysis is the dominant in this research. As already stated, open ended interview questionnaire was used for collecting primary data. Notably, various types of questionnaires were prepared to collect information because of diversity of the respondents.

### *1.5.1 Document Evaluation*

Identifying the lack of information is an essential and indispensable part of research. To ascertain the credibility of this research, the author conducted a comprehensive examination of pertinent literature, including books, journals, research papers, and online resources, with the aim of identifying any areas of insufficient knowledge. The author covers some studies that are relevant to this subject.

Secondary sources were used in order to analyze the existing literature. A number of public documents pertaining to environmental policy, legislation, planning, and strategies were reviewed and analyzed for key term definition and develop the relation with all variables. In addition to relevant books, journal articles and reports, the reviewed documents include the National Action Plan (2022–2042) for the implementation of the national strategy on internal displacement, published by the MODMR (2022), livelihood pattern analysis of IDPs due to the effects of natural

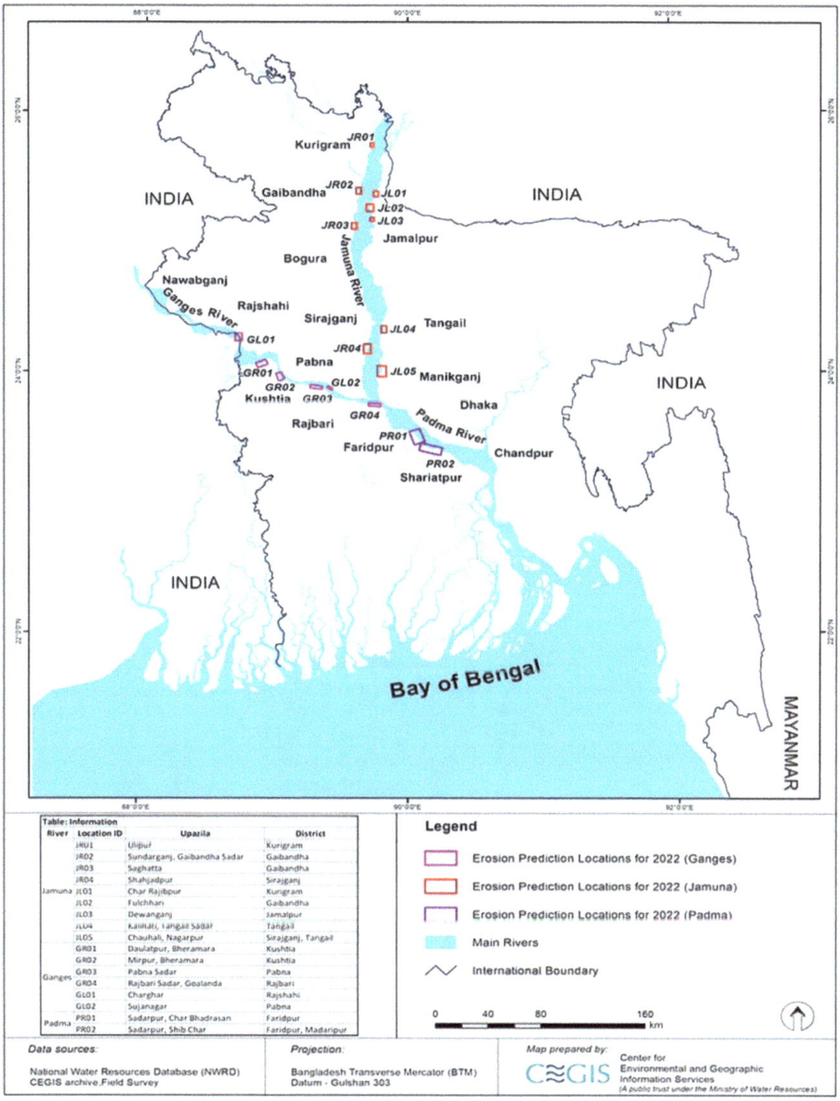


**Fig. 1.3** Major rivers and erosion prone locations in Bangladesh

disaster, including river ban erosion, conducted by CEGIS (2010), and supported by Comprehensive Disaster Management Programme (CDMP 2014). Environmental Conservation Act of 1995, and the Environmental Policy of 1992. These documents were examined to determine the sufficiency and precision of vulnerability analysis and river bank erosion. In addition, the library of the Institute of Environmental science (IES) at the University of Rajshahi, the archives of Ministry of Environment, Forest and Climate Change (MOEFCC) and various report of BWDB, DPHE were consulted as secondary sources of data. However, it is now common practice for secondary sources to be gathered through Internet research.

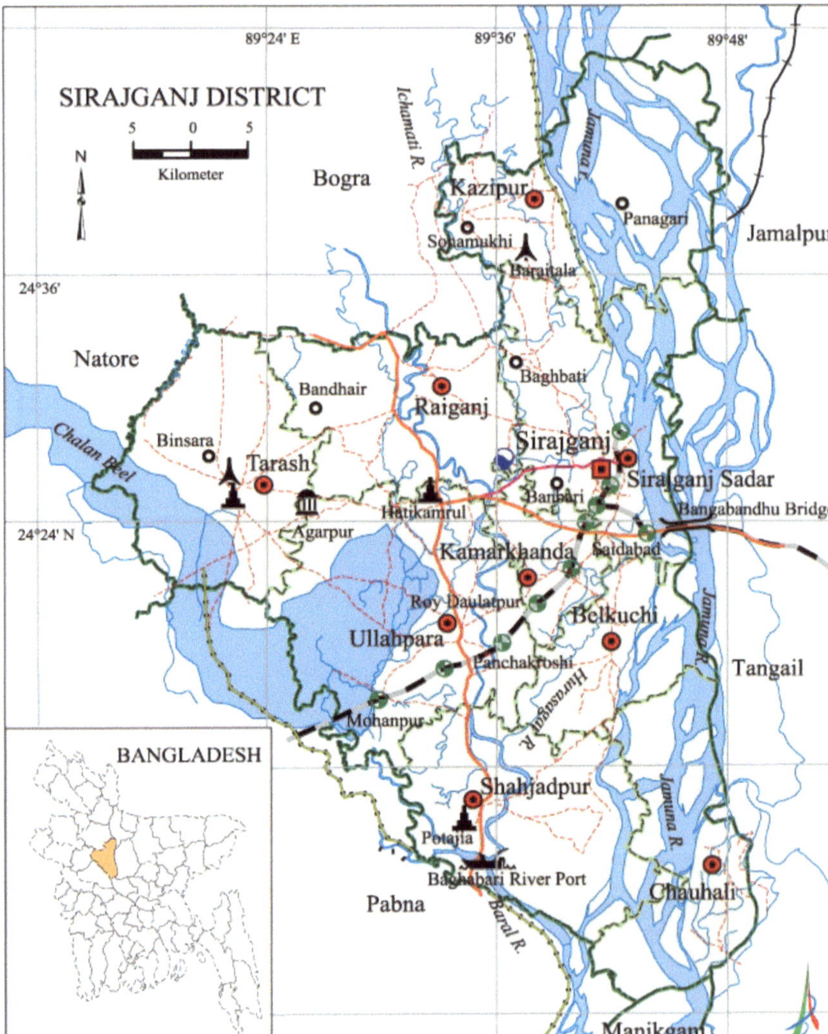


**Fig. 1.4** Sirajganj District, Bangladesh (Banglapedia 2023a)

### 1.5.2 *Data Analysis*

The primary purpose of data analysis is to explore the relationship between the dependent and independent variables being studied. Creswell mentioned that there is no single way of analyzing data, and that analysis is a process which attempts to make sense of data. After collecting field data, primary data were prepared for descriptive analysis, and supplemented with quantitative data, on a limited scale.

**Fig. 1.5** Map of Shahjadpur Upazila, Bangladesh (Banglapedia 2023b)

During data analysis the author continually reflected on field data, moving into further understanding and representation of data, and providing an interpretation of data. After the fieldwork, the interviews were translated from Bengali (local dialect in Sirajganj) to English. The texts from interviews, focus group discussions (FGDs), and key informant interviews (KIIs) were thoroughly examined. This technique involved breaking down the data into relevant concepts and describing those using concise sequences of words.

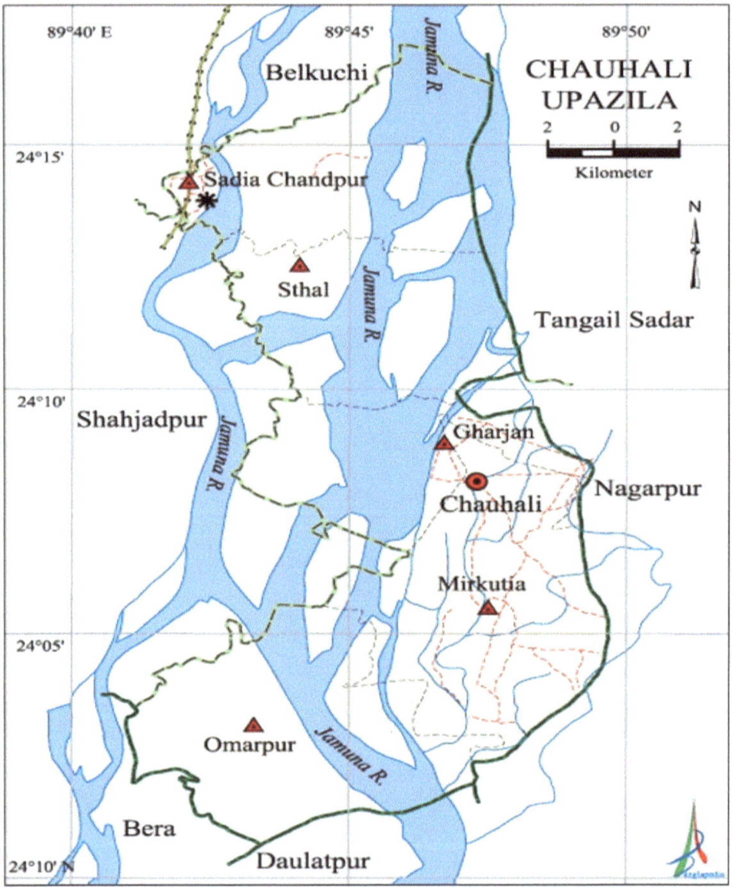


**Fig. 1.6** Map of Chauhali Upazila, Bangladesh (Banglapedia 2023c)

## 1.6 Conclusion

This chapter presented a summary of the research study, including its nature and an extensive overview of how the author conducted the research. The subsequent chapter explores the theoretical framework and associated concepts used in this study, and presents an overview of internally displaced persons (IDPs) affected by natural disasters, as well as the factors that contribute to their vulnerability, as documented in the published literature. The subsequent chapter delves into the researcher's examination of the various factors that exerted effect on migration and subsequently impacted livelihood of the affected individuals. The study's further findings are categorized into overarching thematic groups. The conclusions of the study include a contextual analysis and recommendations, which are presented in Chap. 6.

# References


Abrar CR, Azad SN (2007) Coping with riverbank erosion induced displacement, RMRRU. Available at https://www.rmmru.org/newsite/wp-content/uploads/2019/10/Policy-Brief-01.pdf

Adger WN, Arnell NW, Tompkins EL (2015) Successful adaptation to climate change across scales. Glob Environ Chang 15(2):77–86. https://doi.org/10.1016/j.gloenvcha.2004.12.005

Akhter R (1984) Causes and consequences of migration to the city: a case study of slum dwellers in Dhaka. Unpublished manuscript, Centre for Urban Studies, Bangladesh

Alexander D (2002) Principles of emergency planning and management. Oxford University Press, New York

Baki Abdullahel ATM (2014) Socio-economic impacts of Gorai riverbank erosion on people: a case study of Kumarkhali, Kushtia. Doctoral dissertation, BRAC University. Available at https://dspace.bracu.ac.bd/xmlui/handle/10361/3532

Banglapedia (2023a) Sirajganj District. Bangladesh National Portal. Available at https://en.banglapedia.org/index.php?title=Sirajganj_District

Banglapedia (2023b) Shahjadpur Upazila. Bangladesh National Portal. Available at https://en.banglapedia.org/index.php/Shahjadpur_Upazila

Banglapedia (2023c) Chauhali Upazila. Bangladesh National Portal. Available at https://en.banglapedia.org/index.php/Chauhali_Upazila

Bhuiyan MAH, Islam SMD-U, Azam G (2017) Exploring impacts and livelihood vulnerability of riverbank erosion hazard among rural household along the river Padma of Bangladesh. Environ Syst Res 6(25). https://doi.org/10.1186/s40068-017-0102-9

Black R (2001) Environmental refugees: myth or reality? New issues in refugee research. University of Sussex, Falmer, Brighton BN1 9SJ, United Kingdom. ISSN 1020-7473. Available at http://www.unhcr.org

Brauch HG, Behera NC, Kameri-Mbote P, Grin J, Spring ÚO, Chourou B, Mesjasz C, Krummenacher H (2009) Facing global environmental change: environmental, human, energy, food, health and water security concepts. Hexagon series on human and environmental security and peace, vol 4. Springer Science & Business Media

Brown O (2008) Migration and climate change. IOM migration research series, paper no 31. Available at: www.iom.int

Bryman A (2012) Social research methods. Oxford University Press, Oxford

CEGIS (2010) Prediction of riverbank erosion along the Jamuna, Ganges and Padma rivers in 2010. North South University Library OPAC. Available at https://opac.northsouth.edu/cgi-bin/koha/opac-detail.pl?biblionumber=23672&shelfbrowse_itemnumber=37605#shelfbrowser

Chuenchum P, Xu M, Tang W (2019) Estimation of soil erosion and sediment yield in the Lancang–Mekong River using the modified revised universal soil loss equation and GIS techniques. Water 12(1):135. https://doi.org/10.3390/w12010135

Comprehensive Disaster Management Programme (CDMP II) (2014) Trend and impact analysis of internal displacement due to the impacts of disaster and climate change. Ministry of Disaster Management and Relief, Dhaka. Available at https://www.undp.org/bangladesh/publications/trend-and-impact-analysis-internal-displacement-due-impacts-disaster-and-climate-change

Dewan A, Corner R, Saleem A, Rahman MM, Haider MR, Rahman MM, Sarker MH (2017) Assessing channel changes of the Ganges–Padma River system in Bangladesh using Landsat and hydrological data. Geomorphology 276:257–279. https://doi.org/10.1016/j.geomorph.2016.10.017

Earl B (1983) The practice of social research, 3rd edn. Wadsworth Publishing Company, Belmont

Eckstein D, Hutfils M, Winges M (2019) Global Climate Risk Index 2019 who suffers most from extreme weather events? Weather-related loss events in 2017 and 1998 to 2017. Germanwatch. Available at https://germanwatch.org/files/Global%20Climate%20Risk%20Index%202019_2.pdf

Elahi KM, Rogge RJ (1991) Riverbank erosion, flood and population, displacements in Bangladesh. In: Elahi KM, Ahmed KS, Mafizuddin M (eds) Riverbank erosion, flood hazard and population displacement in Bangladesh, Dhaka. Riverbank Erosion Impact Study (REIS), p 364

Elahi KM, Ahmed QS, Mafizuddin M (eds) (1991) Riverbank erosion, flood and population displacement in Bangladesh. Riverbank erosion impact study. Jahangirnagar University, Savar, Dhaka

Ercan A, Younis BA (2007) Prediction of flow and bank erosion in the Sacramento River. World environmental and water resources congress 2006: examining the confluence of environmental and water concerns, Omaha, NE, Amerika Birleşik Devletleri. https://doi.org/10.1061/40856(200)352

Etzold B, Ahmed AU, Neelormi S (2014) Clouds gather in the sky, but no rain falls. Vulnerability to rainfall variability and food insecurity in Northern Bangladesh and its effects on migration. Clim Dev 6(1):18–27. https://doi.org/10.1080/17565529.2013.833078

Florsheim JL, Mount JF, Chin A (2008) Bank erosion as a desirable attribute of rivers. BioScience 58(6):519–529. https://doi.org/10.1641/B580608

Government of Bangladesh (2009) Bangladesh climate change strategy and action plan (BCCSAP). Ministry of Environment and Forest, Dhaka, Sept 2009. Available at https://policy.asiapacificenergy.org/sites/default/files/Bangladesh%20Climate%20Change%20Strategy%20and%20Action%20Plan%202009.pdf

Haque CE (1987) Impact of riverbank erosion in Kazipur: an application of Landsat imagery. REIS Newsletters. Jahangirnagar University, Savar, Dhaka, Bangladesh

Hossain MZ (1984) Riverbank erosion and population displacement: a case of Kazipur in Pabna. Unpublished thesis, Jahangirnagar University, Savar, Dhaka

Hossain M (1993) Economic effects of riverbank erosion: some evidence from Bangladesh. Disasters 17(1):25. https://doi.org/10.1111/j.1467-7717.1993.tb00485.x

Hunter LM, Luna JK, Norton RM (2015) Environmental dimensions of migration. Annu Rev Sociol 41:377–397. https://doi.org/10.1146/annurevsoc-073014-112223

IPCC (2007) Climate change 2007: the physical science basis: Contribution of Working Group I to the Fourth Assessment Report of the Intergovernmental Panel on Climate Change. Cambridge University Press, Cambridge; New York. Available at https://www.ipcc.ch/site/assets/uploads/2018/02/ar4-wg1-frontmatter-1.pdf

IPCC (2022). Climate change 2022: impact, adaptation and vulnerabilities. Intergovernmental Panel on Climate Change. Available at https://www.ipcc.ch/report/sixth-assessment-report-working-group-ii/

Islam MK, Kibriya NA, Dustegir MM (2018) Impact analysis of sand dredging from alluvial tidal river. In: E3S web of conferences, vol 40. EDP Sciences, p 03036. https://doi.org/10.1051/e3sconf/20184003036

Kabir ME, Kamruzzaman P (2022) Exploring the drivers of vulnerability among disadvantaged internal migrants in riverbank erosion prone areas in North-West Bangladesh. J South Asian Dev 17(1):57–83. https://doi.org/10.1177/09731741221078741

Kaur SP (2013) Variables in research. IJRRMS 3(4). Available at https://sites.usp.br/rnp/wp-content/uploads/sites/830/2020/10/complemento-aula-1.pdf

Kerlinger FN (1983) Foundation of behavioral research, 2 edn. Holt, Rinehart and Winston, Inc, USA, pp 29–40

Keya MK, Harun SMR (2007) Riverbank erosion induced stress and coping of displaced women in Bangladesh. Empowerment 14:17–30

Mehedi H, Nag AK, Farhana S (2010) Climate induced displacement—case study of Cyclone Aila in the southwest coastal region of Bangladesh. CLEAN, Khulna, July 2010. https://doi.org/10.13140/RG.2.1.3490.9921

Ministry of Disaster Management and Relief (2022) Government of the People's Republic of Bangladesh, National Action Plan (2022–2042) to implement the National Strategy on Internal Displacement Management, Bangladesh Secretariat, Dhaka-1000. Available at https://modmr.portal.gov.bd/sites/default/files/files/modmr.portal.gov.bd/publications/9faf92a8_f054_4a04_a290_91ee14bb88b0/Eng_National%20Action%20Plan%20(2022-2042)%20on%20Internal%20Displacment%20(4).pdf

Momen MN (2021) Multi-stakeholder partnerships in public policy. In: Leal Filho W, Marisa Azul A, Brandli L, Lange Salvia A, Wall T (eds) Partnerships for the goals. Encyclopedia of the UN Sustainable Development Goals. Springer, Cham. https://doi.org/10.1007/978-3-319-95963-4_50

Momen MN, Ferdous J (2023) Public health response and state capacity in Bangladesh: COVID-19 preventive measures. In: Governance in Bangladesh. New frontiers in regional science: Asian perspectives, vol 68. Springer, Singapore. https://doi.org/10.1007/978-981-99-0424-2_5

Muinul I (1992) Natural calamities and environmental refugees in Bangladesh. Refugee 12(1). https://doi.org/10.25071/1920-7336.21639

Myers N, Kent J (1995) Environmental exodus: an emergent crisis in the global arena. Climate Institute of Washington D.C.

Nde SC, Bett SK, Mathuthu M, Palamuleni LG (2022) Anthropogenic land use and land cover change as potential drivers of sediment sources in the upper Crocodile River, North West Province, South Africa. Int J Environ Res Public Health 19(20):13313. https://doi.org/10.3390/ijerph192013313

NIRAPAD (2021) Monthly hazard incidence report, June 2021, Bangladesh. Available at https://www.nirapad.org.bd/home/resources/monthlyHazard

Okeke CAU, Ede AN (2019) Mechanisms of riverbank failure and channel instability on the Nkisi River, Southeast Nigeria. IOP Conf Ser Mater Sci Eng 640:012104. https://doi.org/10.1088/1757-899x/640/1/012104

Okeke CAU, Azuh DE, Ogbuagu FU, Kogure T (2020) Assessment of land use impact and seepage erosion contributions to seasonal variations in riverbank stability: the Iju River, SW Nigeria. Groundwater Sustain Dev 11:100448. https://doi.org/10.1016/j.gsd.2020.100448

Patel HK, Kumar B (2023) Hydro-morphological behavior around T-shaped spur dikes with downward seepage. Sci Rep 13(1). https://doi.org/10.1038/s41598-023-37694-w

Patwary OH (2016) Climate change, human displacement and the potentiality of conflict: the case of Bangladesh. Masters thesis, UiT Munin. Available at https://munin.uit.no/handle/10037/22675

Piguet E (2013) From 'primitive migration' to 'climate refugees': the curious fate of the natural environment in migration studies. Ann Assoc Am Geogr 103(1):148–162. https://doi.org/10.1080/00045608.2012.696233

Rabbi H, Saifullah ASM, Sheikh MS, Sarker MMH, Bhowmick AC (2013) Recent study on river bank erosion and its impacts on land displaced people in Sirajgonj Riverine Area of Bangladesh. World J Appl Environ Chem Int J Appl Environ Sci 2(2):36–43

Rahman SH, Islam SMD (2016) Degrading riverine ecology of Bangladesh and options for management. SUB J Sustain Env Dev 1(2016):11–27

Rana MS, Nessa AM (2017) Impact of riverbank erosion on population migration and resettlement of Bangladesh. Sci J Appl Math Stat 5(2):60–69. https://doi.org/10.11648/j.sjams.20170502.11

Rasheed KBS (2008) Bangladesh-resource and environmental profile. AHD Publishing House, Dhaka, Bangladesh, p 257

Terminski B (2012) Towards recognition and protection of forced environmental migrants in the public international law; Refugee or IDPs umbrella? Working paper. Social Science Open Access Repository (SSOAR), p 21. Available at https://nbn-resolving.org/urn:nbn:de:0168-ssoar-329056

Uddin AFMA, Basak JK (2012) Effects of riverbank erosion on livelihood. Unnayan Onneshan—The Innovators, Dhaka, Bangladesh, Jan 2012. https://doi.org/10.13140/RG.2.2.24933.37609/1
Udomratn P (2008) Mental health and the psychosocial consequences of natural disasters in Asia. Int Rev Psychiatry 20(5):441–444. https://doi.org/10.1080/09540260802397487
UWICED (2002) Vulnerability and small island developing states, prepared for the UNDP/CAPACITY 21 programme of action for the sustainable development of small island states. University of the West Indies Centre for Environment and Development. Available at http://www.sopac.org.fj/Projects/Evi/Files/Env%20vuln%20of%20SIDS.pdf
Walsham M (2010) Assessing the evidence: environment, climate change and migration in Bangladesh. International Organization for Migration (IOM). Available at https://publications.iom.int/system/files/pdf/environment_climate_change_bangladesh.pdf

# Chapter 2
# Riverbank Erosion: Operationalizing the Risks and Vulnerability

**Abstract** Riverbank erosion is a multifaceted and dynamic phenomenon that presents considerable hazards to communities and infrastructure adjacent to river bank. This chapter defines and applies the main ideas of the theoretical investigation and examines in a critical manner about the patterns of livelihood risks and vulnerabilities of Internally Displaced Persons (IDPs) caused by riverbank erosion in Bangladesh. This chapter consists of a thorough literature review that is relevant to the research context, and primarily aims to identify and discuss essential concepts used in the study. The chapter also examines the theoretical models and frameworks that are closely linked, and aims to establish a conceptual framework that may serve as the guiding principle for the current study and provide an explanation of the framework.

## 2.1 Introduction

Riverbank erosion is a critical environmental concern that, presents substantial problems to communities, infrastructure, and ecosystems (Dekaraja and Mahanta 2021). In the tropical and subtropical regions of Asia and Africa, intensified riverbank erosion has been associated with increased anthropogenic pressures resulting from swift alterations in agricultural and riparian land uses (Okeke et al. 2020). River bank erosion, a worldwide occurrence, resulting in a deterioration of ecosystem services and functions.

The tropics and subtropics are especially susceptible, with Africa and Asia experiencing some of the most significant repercussions on food security. In Ethiopia, river bank erosion has resulted in significant environmental and social consequences, particularly in the Beles watershed of the Upper Blue Nile Basin. Worldwide, approximately 80% of agricultural land degradation results from river



M. Begum, *Understanding River Bank Erosion in Bangladesh*, Springer Water,
https://doi.org/10.1007/978-3-032-00288-4_2

bank erosion, leading to an annual loss of 10 million hectares of arable land, incurring a cost of $523.1 million each year. Africa is the continent most impacted, housing 50% of the global population affected by river bank erosion. In sub-Saharan Africa, 77% of the area is impacted by river bank erosion, caused by population pressure, the exploitation of fragile ecosystems, and overgrazing (Kebede et al. 2020). Soil erosion is a significant contributor to riverbank instability, resulting in undercutting and mass failure (Thi and Minh 2019). Riverbanks are a crucial element of riparian zones, offering habitat and affecting the general functionality of these essential ecosystems (Florsheim et al. 2008).

Effective management of riverbank erosion necessitates a comprehensive, risk-informed strategy that takes into account the complex relationships among the soil-sediment-water system and the diverse economic activities in the river basin (Brils et al. 2013). Fluctuations in river water levels, especially during the rainy season, can increase water shear stress and intensify soil erosion, leading to the development of overhanging riverbanks (Thi and Minh 2019). Understanding and measuring the dangers linked to riverbank erosion, together with the possible effects of climate change, is essential for formulating efficient flood control infrastructure and management plans.

Riverbank erosion further exacerbates hydrological threats, including flooding, surface water contamination, and rainfall-triggered slope failures. The global urban population is expected to increase by 2.5 billion by 2050, with Asia and Africa accounting for 90% of this growth. Consequently, the pressure on riverine ecosystems are likely to escalate, requiring an in-depth comprehension of riverbank erosion mechanisms and efficient mitigation approaches.

Riverbank erosion becomes a major natural hazards in Bangladesh. This leads to substantial damages of the riverbed. Each year, erosion affects millions of Americans, leading to the depletion of various resources. Specifically, almost 5% of the country's whole floodplain is directly impacted by this form of erosion (Munna and Nayem 2018). The phenomenon has also both direct and indirect impacts on people's lives and the economic resilience (Bhuiyan et al. 2017). Riverbank erosion is considered a significant barrier to the country's achievement of its SDGs 2, particularly those relating to eliminating hunger and poverty (Momen and Ferdous 2023). Riverbank erosion in Bangladesh is considered a significant and steady factor contributing to the loss of property (Shamsuddoha and Chowdhury 2007). Climate change has recently led to rapid erosion of riverbanks, resulting in the collapse of Unions and Upazilas. Riverbank erosion now becomes a frequent natural disaster that has enduring consequences (Das et al. 2014).

Riverbank erosion refers to the sudden manifestation of soil degradation along the river, which can be induced by natural or human factors (Rahman et al. 2015). It can be either voluntarily or compelled, and may result in temporary or permanent displacement. Put simply, it is a prevalent factor that leads to internal forced migration and relocation within Bangladesh. However, people living along the country's rivers and on river bank, known as 'chars', face a persistent danger from riverbank erosion. Riverbank erosion is attributed to deforestation, flooding, intense precipitation, the powerful flow of the river, and the deposition of silt. Since 1973, a total of 158,780 hectares of land have gone through erosion. Only in the year 2010, 11,000 residents living along the Jamuna riverside and over 5000 residents living along the Ganges and Padma riverbanks were displaced by river bank erosion (CEGIS 2010). River bank erosion is influenced by various factors, such as changes in shear strength, alterations in the river water course, nature of erosion-prone bank, obstructions in the streams, and wind waves (Rahman et al. 2015). The National Strategy on Internal Displacement Management, as defined by the GOB, refers to individuals, groups, households, or the entire communities who have been compelled to abandon their ancestral homes either temporarily or permanently caused by quick or gradual changes in climatic conditions and processes. MODMR (2021) notes that these displaced individuals have not crossed the recognized state border.

Islam and Rashid (2011) note that riverbank erosion is a unique natural disaster, surpassing all other environmental calamities. During every disaster, individuals often lose their infrastructure, and affected households not only lose their ancestral shelter but also their assets, leaving them homeless. The only distinction is that the number of deaths is lower. The displacement caused by riverbank erosion is extensive. There are certain losses that are invaluable and cannot be monetized. In addition to the loss of land, they also experience various forms of deprivation. To effectively encounter the conceptual confusion, the classification of displaced individuals is spelt out in the CDMP-II on displacement. This classification categorizes internally displaced individuals into three distinct groups: those who are temporarily displaced, those who are in-between temporal and permanent displacement, and those who are permanently displaced (MODMR 2021). However, riverbank erosion in Bangladesh is a major natural disaster that disturbs the balance of rivers and coastal habitats by gradually eroding the river bank materials (Nishat and Mukherjee 2013). The geomorphic process described here refers to the movement of the shoreline from where it started (Islam and Rashid 2011).

## 2.2 Internally Displaced Persons (IDPs)

The interconnections between climate change, disasters, and IDPs are intertwined, although there is the absence of agreement on the terminology and ideas used to describe individuals who migrate due to river bank erosion. The existing literature is filled with many terms, for instance, 'environmental migrants', 'climate refugees', 'environmental refugees', and 'environmentally displaced persons', among others. However, rather than clarifying the subject, these terms just further to its complexity. The impacts of climate change on displacement is observed in the human displacement, which presents significant difficulties to legislation, and institutions. The environmental displacement caused by climate change is undeniably a new reality that requires new and innovative responses in the academic, and political-institutional domains. On the other hand, as per the guiding principles of the UNCHR, IDPs are individuals or groups who have been forced to leave or abandon their original of residence due to armed conflict, widespread civil war and violence, widespread human rights violation, natural or artificial disasters, and have not traversed the national territory (IDMC 2011).

The UN Environmental Program (UNEP) expanded the definition of refugees to include individuals who have been migrated due to environmental factors. Environmental refugees are individuals who have to flee their usual living environment, either temporarily or permanently, because of the environmental disturbances caused by natural factors and/or human activities, which pose a threat to their survival and significantly impact their quality of life (IDMC 2011).

The IOM in 2007 introduced the term "environmental migrants" to describe individuals or groups who are compelled to leave their home of origin due to sudden or gradual changes in the environment that negatively impact their lives or livelihood. These individuals may be forced to move temporarily or permanently, either within their own country or to another country. The terms 'Environmental IDP' are used synonymously. The term "displaced persons" refers to individuals or groups who have been compelled to leave home of origin for an extended period of time as a result of natural or man-made calamities (Anastasiou and Ghráinne 2018). Internally displaced persons (IDPs) refer to those who are unable or unwilling to leave their country and are thus compelled to relocate within their own country (Ferris 2011). IDPs remain within their country due to several factors, including inability to undertake long-distance journeys, and remote and inaccessible border areas. IDPs often experience acute scarcities of food, water, and other essential necessities, and often get minimal physical security. International law provides minimal safeguards for them, and widely accepted agreements among the signatory countries (such as the Geneva Convention, 1949) are often difficult to put into practice. After the identification of an IDP crisis, foreign response and development partner organizations may encounter difficulties and potential dangers in obtaining access (Moore and Shellman 2002).

The extent to which internal migrants are susceptible to riverbank erosion has not been thoroughly investigated. Empirical evidences from two riverbank erosion-prone districts in Bangladesh reveals that affected family members face several risks and vulnerabilities related to economic, institutional, environmental, and social factors. Socioeconomic systems, such as societal governance and income distribution, along with environmental change, contribute to a multitude of linked vulnerabilities. It is worth noting that environmental change is not the primary factor influencing human wellbeing (Adger and Brown 2009). In recent years, the international community has neglected to recognize IDPs. Nonetheless, it represents a significant and pressing issue in contemporary society, as highlighted by Anastasiou and Ghráinne (2018). As per the IDMC (2011) assessment report, internal displacement causes immense hardship for millions of individuals, contributes to food insecurity, and hinders long-term development. Throughout the process of being forced to leave their homes, IDPs face insurmountable obstacles. The individuals experienced marginalization due to being displaced, which affected their means of making a living and caused psychological and physical distress.

## 2.3 Environmentally Displaced Persons (EDPs)

The UNHCR and IOM have recognized a distinct group of individuals known as Environmentally Displaced Persons (EDPs), who are forced to leave home of origin due to environmental factors and resultant climate change. This acknowledges the existence of a growing and significant population of individuals who are forcibly displaced, but do not fall under the official categorizations of refugees or IDPs. EDPs are those who are displaced either within their own national territory or across international borders due to significant environmental degradation, deterioration, or destruction. While environmental factors are a key reason for their displacement, they may not be the only cause (IOM 2007). According to David, environmental displacement refers to the situation where individuals are compelled to migrate initially as a result of environmental impacts.

Riverbank erosion poses a substantial and persistent concern in Bangladesh, a nation located in the Bengal delta and traversed by three principal rivers, the Ganges, the Brahmaputra, and the Meghna. Bhuiyan et al. (2017). This natural, gradual danger persistently erodes bank materials, disrupting the ecological equilibrium and displacing communities that have resided along these rivers for millennia (Mallick and Mallick 2021). However, IDPs are an undeniable consequence of social changes occurring in both the regions where displacement occurs and the regions that receive these displaced individuals. The situation is highly particular, the inclusion of a new member into a familiar or unfamiliar community. Put it simply, social exclusion pressure on the original community and social inclusion put pressure on the new community.

### *2.3.1 Social Exclusion*

Social exclusion is the socio-economic situations that lead to a loss of both human capital and has a definite impact on the overall integration of social patterns. The displaced individuals are prepared to surrender their initial state of livelihood. Social exclusion, under certain situations, appears to be a significant factor contributing to a diverse array of social and economic issues (Nishat and Mukherjee 2013). Furthermore, social exclusion gives rise to the concept of social inclusion. On one hand, those migrating to the new areas may confront various familiar or unfamiliar barriers. Occasionally, certain events have a profound and lasting effects on human life. While some households are able to overcome and go forward, others continue to suffer for the remainder of their lives (Nishat and Mukherjee 2013). Social exclusion entails the displacement of IDPs and the subsequent transfer of their economic resource.

Haque and Zaman (1989) propose that individuals living along river banks perceive erosion as an unavoidable. Hence, households that experience the loss of their homes or farmland as a result of erosion decide to stay in the nearby vicinity.

According to a survey conducted on 547 families living near the Jamuna River, it was found that the average distance households traveled was 2.2 km. Further, less than 3% of households moved a distance greater than 8 km. A study conducted in the northwestern part of Bangladesh examined 200 households that were affected by displacement. The study revealed that, on average, these households had experienced displacement 4.46 times (Abrar and Azad 2007).

Urban areas benefit greatly from the availability of inexpensive human resources due to the presence of IDPs. Migrated people are usually felt uncertain about the new employment arrangements, as occupational shifts necessitated the acquisition of new skills. Occasionally, the working conditions and may be unpleasant. Occasionally, a small number of migrants earned more incomes than they did in their original communities. Furthermore, individuals who formerly occupied influential roles within rural communities may occasionally subject to marginalization when they were compelled to undertake less prestigious occupations in urban areas. However, many families experienced the loss of property and land because of river bank erosion, certain individuals may make the choice to refrain from migrating. They preferred addressing hazards as they occurred (Abrar and Azad 2007).

### 2.3.2 *Social Inclusion*

Social inclusion is such a strategy between the migrants and the residents who are receiving each other with maximum adaptation. Social inclusion means how much they are capable to include themselves in the process of social transformation in one side and from the other side how much the traditional people are flexible to adjust with the new comer. But the issue is very challenging due to social and cultural traditions of the displaced households and from the view of traditional people they are quite afraid.

Internal movements typically arise from the persistent links between two local areas, which are influenced by political, trade, investment, educational, and cultural factors. Migratory movements, once initiated, evolve into self-perpetuating societal phenomena. Undoubtedly, the arrival of a new individual brings about significant changes in their way of life. This particular form of migration further exacerbates pre-existing ethnic or racial divides in civilizations that have historically had minority populations. However, migration is generally believed to result in cultural transformations, which might be seen as potentially alarming.

Families possessing significant land assets have the opportunity to move to an unfamiliar area, whilst those with little means must depend on familial and social connections for the purpose of resettlement. Although displacement can have severe consequences for many households, including a significant decrease in assets and means of survival, as well as social and mental effects of forced displacement (Haque and Zaman 1989).

Whatever the process of social inclusion, in reality the IDPs are engaged with their regional people for their comfortable behavior and livelihoods and on the other side the original residents are not enough curious for the migrants. The newcomer generally fail to attain the educational facilities and they are always afraid about their own religious reflection. They adjust for some years in abroad and then come back to their original home (Haque and Zaman 1989).

Nevertheless, whereas most households predominantly relocate within a certain geographical region, certain households and individuals choose to move temporarily or permanently over longer distances. According to a study conducted by Lein (2009), rural households relied on labor migration within their rural areas during harvesting and sowing seasons, and migration to urban areas for long-term work opportunities for male family members. These migration patterns were seen as crucial strategies for the displaced households to sustain their livelihoods.

## 2.4 Operationalization of Vulnerability and Its Indicators

Vulnerability refers to the extent to which a community possesses the ability to withstand and cope with the adverse effects of climate change. Further, vulnerability refers to the extent to which the environment, economy, or social system can be harmed or degraded by external factors. Vulnerability, originating from the Latin word vulnerabilis, that signifies "to wound," is a quantification of the likelihood for place, individual, group, community, country, or any other entity to experience the effects of disaster. This notion is derived from a synthesis of several physical, social, economic, and environmental aspects (Alexander 2000).

The IPCC (2001) states that vulnerability is measured by the magnitude of the change that a system experiences, its sensitivity to that change, and its ability to cope with the system or process. Vulnerability refers to the state of being exposed to uninsured risks that result in a level of well-being that is considered socially unacceptable (Hoogeveen et al. 2004).

The idea of vulnerability has emerged relatively recently in the literature on disaster management. It is a topic that is open to several interpretations and has significant implications, as discussed by Birkmann (2005). The literature review conducted by León and Carlos (2006) has identified multiple interpretations of the term "vulnerability". Firstly, it is seen as a specific condition of a system prior to a disaster, characterized by factors such as susceptibility, limitations, incapacities, or deficiencies. Further, it refers to the system's inability to encounter the effects of an event (resistance) or to effectively deal with it (coping capacities). Secondly, vulnerability is viewed as a consequence of exposure to a disaster. Lastly, it is understood as the likelihood or possibility of a system experiencing certain outcomes when exposed to a disaster event. These outcomes can be measured in terms of potential damages, including victims or economic losses, or as the possibility of an affected individual surpassing a specific benchmark (León and Carlos 2006).

Vulnerability relates to an individual or group in terms of their ability to deal with, and reconstruct from the effects of disaster (Keya and Harun 2007). Hence, vulnerability is used to describe a society's or community's position in relation to an external threat or hazard. A multifaceted term, vulnerability to effects includes institutional, sociocultural, economic, and bio-geophysical elements. Moreover, vulnerability is a consequence when it is operationalized as the capacity to withstand a disaster event (Islam and Bhuiyan 2016).

It entails a combination of factors that define the extent to which a person's lives are threatened by a distinct and observable event. It refers to exposure to uncertainty and tension, as well as the inability to cope with them. It is worth to consider the numerous social, physical, environmental, economic, and political factors that impact on person, community, or system's vulnerability (Turner et al. 2003; Cannon et al. 2003; UNDRR 2017). It is typically thought to be a result of a system's or process capacity to handle disaster shocks (Smith 1992). Vulnerability can be reduced by taking measures that reduce the likelihood of suffering harm or increased by taking measures that enhance the likelihood of suffering harm (Alexander 2002).

The degree of vulnerability varies with respect to time and space, societal interests, and perceived and actual dangers in diverse academic and policy areas. Even threats that are alarming to one community may not be to another. The survival imperative, which alters priorities in response to space, time, and the intensity of occurrences or phenomena, is one component of security and risk that is universal. Hence, operationalization, criteria, and indicators for the measuring of vulnerability are not widely agreed upon (Brauch et al. 2009). Despite the fact that some researchers have encouraged for a more comprehensive approach that analyzes disaster events in light of the different levels of vulnerability present at households or in the community (Hewitt and Burton 1971; Adger et al. 2005). The author identified and adopted some common form of vulnerabilities of people including individual, communal, social and environmental.

The following Fig. 2.1 shows the domains of vulnerability.

The underlying causes in each instance may be either natural or manmade, or both, and they may change through time and space. The risk of hazards happening, intrinsic resilience (the inherent capacities of a country that perform to deal with natural and human-induced disasters, and resilience all play a role in the three forms of vulnerability. The risks that countries experience due to external shocks to the systems of production, distribution, and consumption depending on economic vulnerability (UWICED 2002). On the other hand, environmental vulnerability refers to the possibility of harm related to the location's ecosystems or environment as well as to physical and biological processes, and ecological resilience (UWICED 2002). Hence, the environmental vulnerability component is the most important of the exposure, sensitivity, adaptive capacity of the system or process.

The majority of specialists agree that analyzing the direct effects of a hazard is insufficient to comprehend vulnerability. Vulnerability refers to the broader environmental and social conditions that limit the capacity of individuals and communities to withstand the effects of a hazard (Schneider 2007). Monitoring vulnerability through time and location, determining the processes that lead to vulnerability, highlighting efforts to reduce it, and assessing the efficacy of these tactics in various social and ecological situations are all possible uses for vulnerability indicators

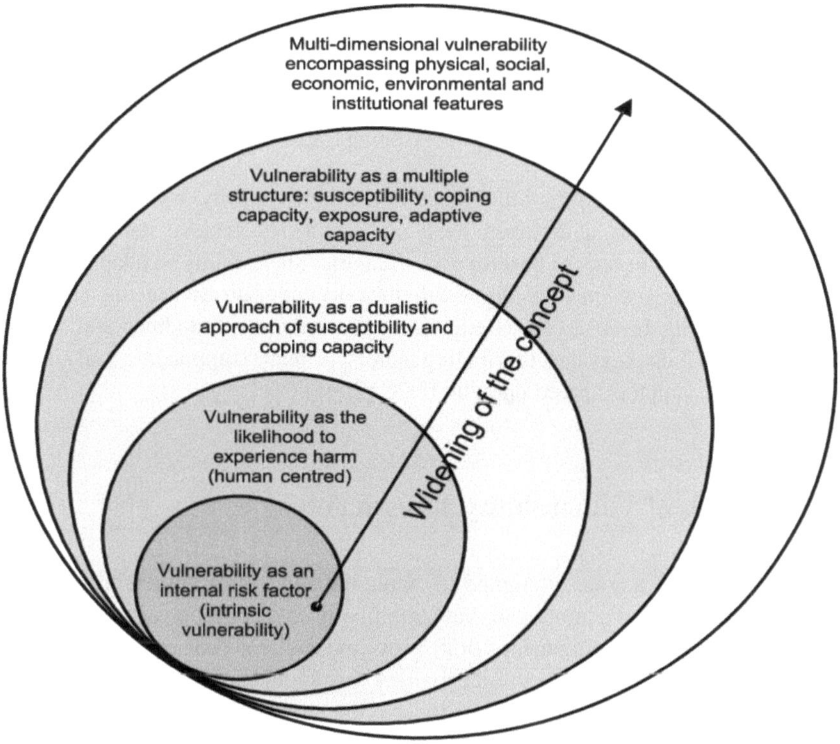


**Fig. 2.1** The spheres of vulnerability (adapted from Birkmann 2005)

(Adger and Brown 2009). A helpful typology is provided by the IPCC (2001), which contends that vulnerability is defined as a function of three contributing components, namely adaptive capacity, sensitivity, and exposure;

$$\text{Vulnerability} = f\left(\text{Exposure}, \text{Sensitivity}, \text{Adaptive Capacity}\right)$$

The views of communities played a significant effect on adaptive capacity, which is what exposure might be described as vulnerability (Frank et al. 2011). The degree to which the area is exposed to a risk was used to determine exposure. It has to do with how much pressure is placed on a specific analytical unit (Islam and Uddin 2015). It can be portrayed as either long-term shifts in severe event frequency and amplitude. Sensitivity is the extent to which a system reacts either favorably or unfavorably to any threat. Economic influence, social connectivity and communication, education, and aptitude to gauge capacity for adaptation in the research area (Islam and Bhuiyan 2016). A system's capacity to adapt to current or future pressures or to cope with their effects is referred to as its adaptive capacity (Islam and Bhuiyan 2016).

White et al. note that a significant argument has been placed in by the Disaster Reduction Institute (DRI) in its report to the DFID regarding vulnerability. Leon conceptualized in the below manner as a synthesis of susceptibility, exposure, and coping capacity.

$$\text{Vulnerability} = \frac{\text{Exposure} \times \text{Susceptibility}}{\text{Coping Capacity}}$$

Twigg (2007) highlighted the importance of adaptive capacity, which refers to the actions and processes undertaken by a community to achieve resilience. This includes strategies to reduce hazards and strengthen their ability to adapt to potential issues. Twigg's perspective aligns with the concept of disaster resilience, where adaptive capacity is observed as equivalent to resilience (Handmer and Dovers 2009). Table 2.1 displays the contributing factors, primary components, and signs of susceptibility, which is agreed upon by IPCC (2001).

## 2.5 Models of Vulnerability Assessment

Various renowned scholar attempted to define the concept of vulnerability and its assessment. Chamber's model of vulnerability is the concepts on vulnerability which leads many new dimensions of it. Watts and Bohle's ideas on vulnerability is also described here in where significantly describe about exposure aspects and coping strategies. On the other hand, Alexander is the leading thinker who talked about the in-depth meaning of vulnerability concept with the close relation of academic knowledge and its practices. Pelling's views also discussed here for better realization of the concepts in order to guide the research.

### *2.5.1 Chamber's Model of Vulnerability*

Chambers (1989) proposed a more methodical conceptualization of vulnerability that specifically focused on communities and their means of subsistence. According to him, vulnerability relates to the state of being exposed to various pressures and events, and the challenges that certain communities have in dealing with these hazards and its contingencies. Chambers (1989) subsequently proposed two dimensions of vulnerability: *the external,* which pertains to susceptibility to external stresses and disturbances, and *the internal,* which is associated with the inability to manage without suffering adverse effects. Shocks, as used in this context, refer to the effects caused by fast and occasionally unforeseeable occurrences, including but not limited to floods, earthquakes, fires, and epidemics. On the contrary, stresses pertain to persistent, accumulative, and foreseeable pressures such as seasonal scarcities and diminishing resources. Furthermore, Chambers (1989) emphasized that vulnerability and poverty are not synonymous, but rather interconnected. In Chamber's view vulnerability, is an opposite of security. Vulnerability may appear

**Table 2.1** Contributing factors, components and indicators of vulnerability (adapted from IPCC 2001)

| Contributing factors of vulnerability | Components | Indicators |
|---|---|---|
| Exposure | Attributes of riverbank erosion disaster, including its magnitude and frequency | – Riverbank erosion frequency<br>– Human death rate<br>– Percentage of households reporting harm<br>– TOTAL number of impacted individuals<br>– Percentage of households reporting land degradation in the past 30 years<br>– Percentage of households that have suffered displacement in the past 30 years<br>– Average number of other natural calamities occurred in the past 30 years |
| | Health sensitivity | – Average time required to reach the nearest healthcare facility<br>– Percentage of households reporting the presence of at least one member with a chronic illness<br>– Percentage of households in which a family member is diagnosed with a communicable disease<br>– Percentage of households lacking proper sanitation facilities and showing poor hygiene conditions<br>– Inverse of life expectancy |
| Sensitivity | Food and agricultural sensitivity | The average month of inadequate food<br>– Land productivity in inverse<br>– The percentage of HHs that get their water straight from a river or pond |
| | Demographic sensitivity | – Dependency ratio (percentage of individuals under 15 and over 65)<br>– The population density<br>– The average family member in an HHs<br>– The percentage of HHs headed by women |

(continued)

**Table 2.1** (continued)

| Contributing factors of vulnerability | Components | Indicators |
|---|---|---|
| Adaptive capacity | Economic capacity | – The proportion of households with just agriculture as a source of income<br>– The average number of households with debt<br>– The percentage of HHs with a family member employed outside the village in a reasonably developed location<br>– The percentage of HHs with access to financial institutions<br>– Monthly income of the household |
| | Social network and communications | – The percentage of homeowners who have received assistance from a neighbor in the last month<br>– The percentage of homeowners who have assisted and supported their neighbors in the last month<br>– The percentage of homeowners whose family members are connected to any kind of organization<br>– The average time it takes to get to the closest vehicle park<br>– How many HHs own communication gadgets (TV, cellphone, etc.) |
| | Education and skill | The percentage of families with people who possess formal or informal skills, as measured by the education index |

at the livelihood level in relation to assets and individuals' management thereof. While vulnerable in this context, assets such as human capital and labor are essential for enabling individuals to deal with such disruptions and stresses (Chambers 1989).

### 2.5.2 *Watts and Bohle's Ideas on Vulnerability*

Watts and Bohle (1993) developed upon Chambers' concepts in 1993, while retaining the framework of vulnerability's external and interior dimensions. According to their view, it is a multidimensional and multilayered issue that is operationalized by the political, economic, and institutional capacities of individuals at particular periods and locations. The correlation between the external and internal dimensions of vulnerability remains consistent in both Chamber's and Bohle's frameworks (Bohle et al. 1994; Bohle 2001): the external side relates to exposure to risks and vulnerabilities, whereas the internal side concerns coping capabilities (Fig. 2.2).

The ***exposure aspect*** of vulnerability pertains to the susceptibility to hazards. This aspect is impacted by two perspectives; Human-Ecological perspectives that scrutinize the dynamics of populations and their abilities to regulate the

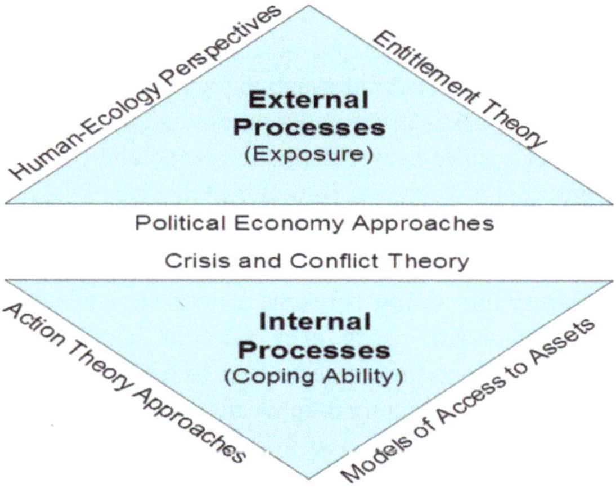


**Fig. 2.2** The two sides of vulnerability (Watts and Bohle 1993)

environment. Entitlement theory, which associates vulnerability with individuals' incapability to acquire or manage resources through lawful economic channels; and political economy approaches, that associate vulnerability with the susceptibility of certain social groups to social inequities (Watts and Bohle 1993; Bohle 2001).

On the other hand, the ***coping aspect*** is impacted by action theory, which examine the methods and channels through which individuals act, whether compelled by economic, societal, or political limitations or out of personal volition. Models of access to asset, which enable individuals to reduce their susceptibility through the acquisition of various types of assets. Crisis and conflict theory, which centers on the management of crisis situations through the control of resources and assets (Watts and Bohle 1993; Bohle 2001).

Bohle's model, similar to the Chambers model, acknowledges the interconnection, such as economic, socio-political, infrastructural, ecological, and personal, and coping capacities. Individuals who possess a greater quantity of assets are less vulnerable to threat, as such assets strengthen their ability to cope with such situations. Bohle notes that the most vulnerable individuals typically have control only over communal assets; thus, the issue of asset control becomes particularly pertinent during times of crisis (Bohle et al. 1994). The model's strength lies in its ability to elucidate not only the expressions of vulnerability, but also its underlying reasons and origins (Watts and Bohle 1993).

### *2.5.3 Alexander's View on Vulnerability*

Vulnerability, according to definition by Alexander (2000), refers to the potential for experiencing casualties, devastation, loss, disturbance, or other forms of damages in connection with a particular disaster. Additionally, Alexander acknowledges that the degree of vulnerability can be increased or decreased based on the course of action in response that the way in which vulnerability is perceived is a crucial consideration. Alexander (2002) establishes a clear nexus between vulnerability and the research undertaken to evaluate it. Based on his thought, vulnerability occurs when research findings are not utilized or disseminated, whereas wilful vulnerability develops when such information regarding vulnerability is willfully disregarded. In addition, Alexander remarks on this matter, stating that an absence of experience with dangers gives rise to pristine vulnerability. These perspectives would result in a proposal of vulnerability that is indicative of the inadequate utilization of research data. On the contrary, he highlights the concept of primary vulnerability, which pertains to an elevated susceptibility to catastrophic harm. On the other hand, secondary vulnerability concerns a deficiency of suitable capabilities and resources that may result in an inadequate or insufficient response during a calamity. Vulnerability in this context refers to weaknesses or incapability's, as opposed to a failure to apply acquired knowledge Alexander (2000). The ISDR has suggested a straightforward method to addressing vulnerability. It operationalizes vulnerability as the combination of physical, social, economic, and environmental issues that make a society more prone to the negative effects of disasters (ISDR 2004).

***Physical factors*** include locational and built-environment vulnerabilities, which are indicated by population density, the isolation of a settlement, its geographical position, and the materials and techniques employed in constructing infrastructure.

Secondly, ***social factors*** encompassing aspects such as gender, healthcare, education, safety and security, protection of human rights, traditional values and beliefs, and organizational structures are examples of social factors.

Thirdly, ***economic factors*** refer to concerns surrounding poverty and encompass variables such as debt levels, degrees of access to financial institutions, insurance, and loans, as well as economic diversity and economic reserves at the individual, community, and national levels.

Lastly, ***environmental factors*** consist of the degradation of natural resources. Poor access to safe air, water, and sanitation, exposure to harmful and hazardous pollutants, and improper waste management practices are all factors that can contribute to environmental vulnerability.

### 2.5.4 *Vulnerability is Interpreted by Pelling (2003)*

Vulnerability is interpreted by Pelling (2003) as susceptibility to risks and the incapacity to prevent or deal with potential current or future damages. Physical vulnerability is defined as the susceptibility of the physical environment; social vulnerability is defined as the vulnerability that individuals and their respective socio-economic, and political systems experience; and human vulnerability is defined as the combining factors of physical and social vulnerability. Pelling also presented a guiding indicators of vulnerability that influence and interact with one another, where appropriate actions can reduce vulnerability, such as exposure, resistance, and resilience (Pelling 2003). However, based on the all scholars view and perception, the author identified the following influential factors of vulnerability that is thoroughly connected with the present research problem.

## 2.6 Influential Factors of Vulnerability

From the above explanation the author identifies the common characteristics of the influential factors and different areas of vulnerabilities that will elaborately discuss in the following chapters. Today, the facts of IDPs are well recognized and numerous research findings also proved that the diversity and trend of internal displace is increasing day by day (Walsham 2010). The connections between influential factors and IDPs related vulnerabilities are unified. IDPs is not just be explained by income differences between two local areas, but obviously there are some factors involve as negative influencer for some securities. No doubt, Bangladesh suffer most due to weak adaptation capacity and poor governance mechanism for livelihoods as well as it suffer most for environmental hazard due to geographical location (Momen and Begum 2021). So, the environmental factors that influence the decisions to migrate in the wake of a disaster. Community is now concerned about their fundamental rights, especially for safe food, equitable education, health care services and housing settlement. The degree to which households or community are adversely affected by damages and dangers is known as social vulnerability. Accordingly, the author develops a diagram of the close relationships among the variables of this research (Table 2.2).

The following Fig. 2.3 shows the relationship among different patterns of vulnerabilities (Infrastructural/Environmental, Socio-institutional, Economic and Social wellbeing) that will give clear ideas about the patterns of risks and vulnerabilities of IDPs in Bangladesh.

**Table 2.2** Identified factors with common characteristics of vulnerability (The author)

| Influential factors | Identified areas of different patterns of vulnerabilities | Common characteristics |
|---|---|---|
| **Bio physical and environmental factors**<br>Inadequate management of the environment, poor architectural and construction methods, uncontrolled urban development, excessive exploitation of natural resources, degradation of ecosystem services that mitigate risks, climate change, and other related factors | Infrastructural and environmental vulnerability | Biophysical loss, including the loss of lands and properties, crop production |
| **Socio institutional factors**<br>Poverty, marginalization, social exclusion, and discrimination, based on gender, socioeconomic position, disability, and age, among others | Social and institutional vulnerabilities | The inability of a society to withstand shocks |
| **Economic factors**<br>The lack of insurance coverage in the informal sector, the susceptibility of rural livelihoods to risks, dependence on single sectors, and other related factors | Economic vulnerability | People's bad economic conditions, loss of financial capital, and rising cost of living |
| **Social wellbeing factors**<br>Health conditions with psychological stress | Health and wellbeing vulnerability | Physical and mental health problems, psychological stress, cultural phenomenon |

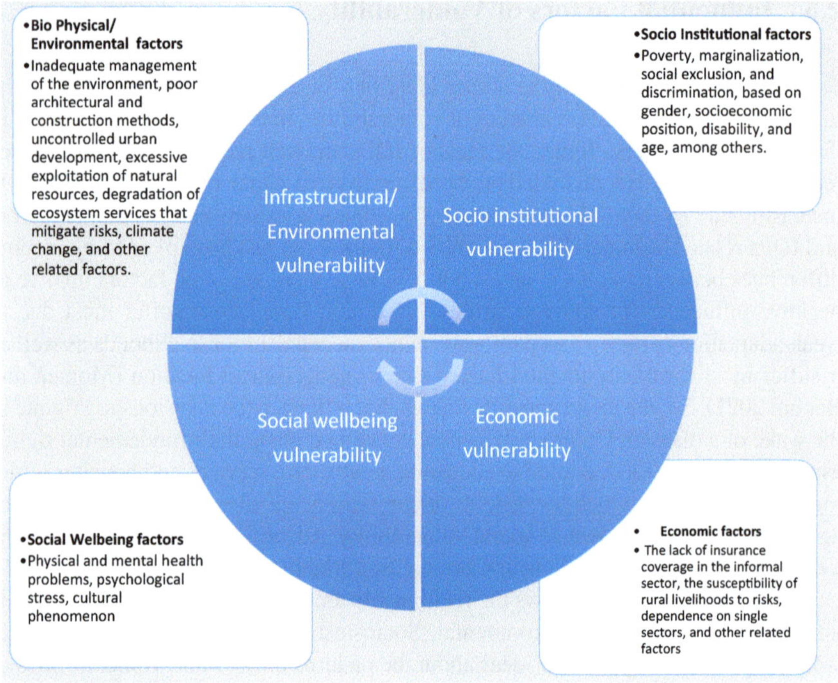


**Fig. 2.3** Influential factors and different pattern of vulnerabilities (The author)

This relationship focuses basically the overall livelihood vulnerabilities of the IDPs because of riverbank erosion in Bangladesh (in Fig. 2.4). Simultaneously the following Chap. 3 will provide the clear explanation of various livelihood vulnerability patterns of IDPs in perspective of the author' assessment through the existing literature and field study.

## 2.7 Conclusion

Various key concepts of risks and vulnerabilities are explained in this chapter. A number of theoretical analysis and frameworks relating to the research context have been discussed briefly as well as their appropriateness to meet the objectives of this study is also explored. This chapter has also analyzed the work of previous researchers in the relevant field, in order to uncover some methodological and knowledge gaps that this research is set to address. The contributing factors, major components and indicators of vulnerability is explained, along with its suitability in addressing the present research issues and concern. The chapter also illustrated the conceptual model that has guided the research context. The conceptual model is explained with

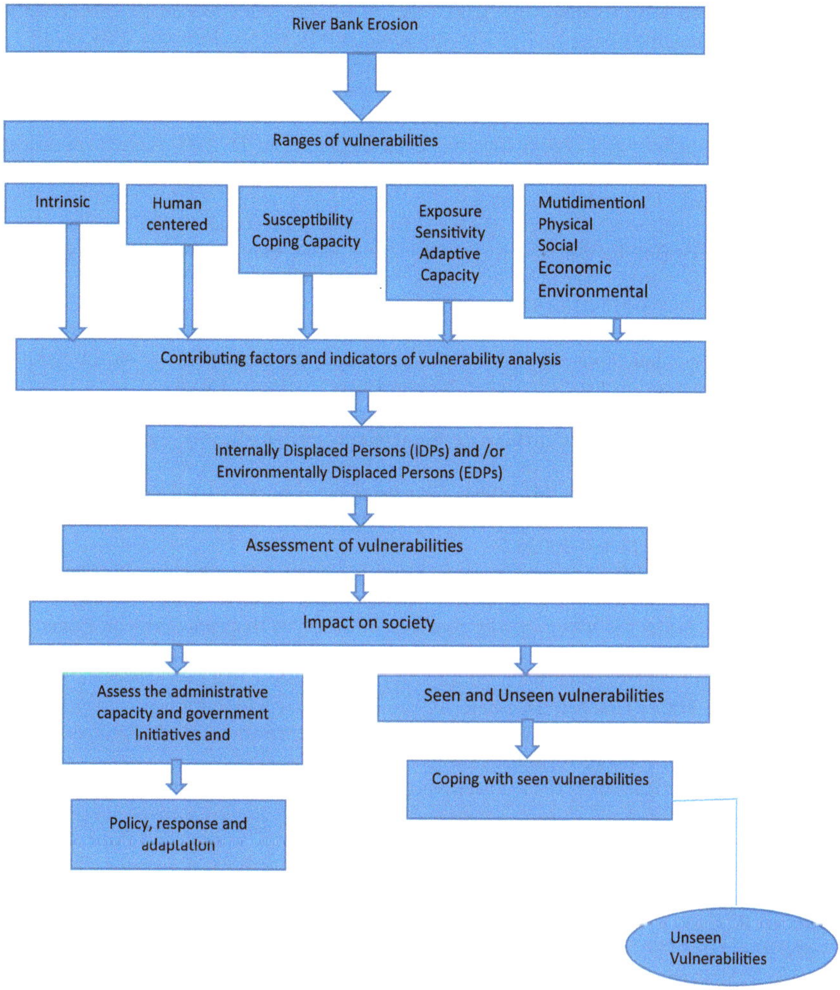


**Fig. 2.4** Conceptual Framework of the study (The author)

a description of the key concepts and how they are related according to the conceptual model illustrated. The key concepts of this study have been operationalized as well with a view to clarify their meaning for the current study.

## References

Abrar CR, Azad SN (2007) Coping with riverbank erosion induced displacement, RMRRU, 2007. Available at https://www.rmmru.org/newsite/wp-content/uploads/2019/10/Policy-Brief-01.pdf

Adger W, Brown K (2009) Vulnerability and resilience to environmental change: ecological and social perspectives. https://doi.org/10.1002/9781444305722.ch8

Adger WN, Arnell NW, Tompkins EL (2005) Successful adaptation to climate change across scales. Glob Environ Chang 15(2):77–86. https://doi.org/10.1016/j.gloenvcha.2004.12.005
Alexander D (2000) Confronting catastrophe. Terra, Hertfordshire
Alexander D (2002) Principles of emergency planning and management. Oxford University Press, New York
Anastasiou T, Ghráinne BN (2018) Environmental displacement in 2018—current protection challenges. RLI working paper mini volume (papers 23–26), refugee law initiative School of Advanced Study, University of London
Bhuiyan MAH, Islam SMD-U, Azam, G (2017) Exploring impacts and livelihood vulnerability of riverbank erosion hazard among rural household along the river Padma of Bangladesh. Environmental systems research. Vol. 6(25). Springer science+business media. https://doi.org/10.1186/s40068-017-0102-9
Birkmann J (2005) Measuring vulnerability to promote disaster-resilient societies: conceptual frameworks and definitions. In: Birkmann J (ed) Measuring vulnerability to natural hazards: towards disaster resilient society. United Nations University Press, Tokyo
Bohle H-G (2001). Vulnerability and criticality. Vulnerability Article 1. Available at https://www.researchgate.net/publication/281020579_Vulnerability_and_Criticality_Perspectives_from_Social_Geography
Bohle HG, Downing TE, Watts MJ (1994) Climate change and social vulnerability: toward a sociology and geography of food insecurity. Glob Environ Chang 4(1):37–48. https://doi.org/10.1016/0959-3780(94)90020-5
Brauch HG, Behera NC, Kameri-Mbote P, Grin J, Spring ÚO, Chourou B, Mesjasz C, Krummenacher H (2009) Facing global environmental change: environmental, human, energy, food, health and water security concepts, Volume 4 of Hexagon Series on Human and Environmental Security and Peace. Springer Science & Business Media 2009
Brils J et al (2013) Introduction: the need for risk-Informed River basin management. In: Brils J, Brack W, Müller-Grabherr D, Négrel P, Vermaat J (eds) Risk-informed Management of European River Basins. The handbook of environmental chemistry, vol 29. Springer, Berlin, Heidelberg. https://doi.org/10.1007/978-3-642-38598-8_1
Cannon T, Twigg J, Rowell J (2003) Social vulnerability, sustainable livelihoods and disasters. DFID, London
CEGIS (2010) Prediction of riverbank erosion along the Jamuna, Ganges and Padma rivers in 2010. North South University Library OPAC Available at https://opac.northsouth.edu/cgi-bin/koha/opac-detail.pl?biblionumber=23672&shelfbrowse_itemnumber=37605#shelfbrowser
Chambers R (1989) Vulnerability. Editorial introduction. IDS Bullet 20(2):1–7, Sussex. https://doi.org/10.1111/j.1759-5436.1989.mp20002001.x
Das KT, Halder HK, Gupta ID (2014) River Bank erosion induced human displacement and its consequences. Liv RevLandscape Res 8(3). https://doi.org/10.12942/lrlr-2014-3
Dekaraja D, Mahanta R (2021) Riverbank erosion and migration inter-linkage: with special focus on Assam, India. Environ Syst Res 10(6). https://doi.org/10.1186/s40068-020-00214-0
Ferris E (2011) Climate change and internal displacement: a contribution to the discussion, UNHCR Bellagio Roundtable, 22–26 February 2011
Florsheim JL, Mount JF, Chin A (2008) Bank erosion as a desirable attribute of rivers. BioScience 58(6):519–529). Oxford University Press. https://doi.org/10.1641/B580608
Frank E, Eakin H, Lopez-Carr D (2011) Social identity, perception and motivation in adaptation to climate risk in the cofee sector of Chiapas. Mexico Global Environ Change 21(1):66–76. https://doi.org/10.1016/j.gloenvcha.2010.11.001
Handmer J, Dovers S (2009) A typology of resilience: rethinking institutions for sustainable development. Earthscan Publications Ltd.. http://hdl.handle.net/1885/24605
Haque CE, Zaman M (1989) Coping with riverbank erosion and displacement in Bangladesh: survival strategies and adjustments. Disasters 13:300–314. https://doi.org/10.1111/j.1467-7717.1989.tb00724.x

Hewitt K, Burton I (1971) The hazardousness of a place: a regional ecology of damaging events. University of Toronto Press
Hoogeveen J, Tesliuc E, Vakis R, Dercon S (2004) A guide to the analysis of risk, vulnerability and vulnerable groups. The World Bank, Washington, DC, USA
IDMC (June 2011) Global estimate 2011: displacement due to natural hazard-induced disasters. Internal Displacement Monitoring Centre (IDMC). Available at https://reliefweb.int/report/world/displacement-due-natural-hazard-induced-disasters-global-estimates-2009-and-2010
IOM (2007) Discussion Note: Migration and the Environment, MC/INF/288, 1 November 2007), Available at www.iom.int/jahia/webdav/shared/shared/mainsite/about_iom/en/council/94/MC_INF_288.pdf p. 1–2
IPCC (2001) Impacts, adaptation and vulnerability. Summary for Policymakers Intergovernmental Panel on Climate Change (Working Group II) 2001. Available at https://www.ipcc.ch/site/assets/uploads/2018/03/WGII_TAR_full_report-2.pdf
ISDR (2004) Living with risk. A Global Review of Disaster Reduction Initiatives (23 November 2004) Available at https://www.unisdr.org/files/657_lwr1.pdf
Islam SMD, Bhuiyan MAH (2016) Exploring climate change adaptation in coastal communities of Bangladesh using adaptation capability index (ACI). Jahangirnagar Univ Environ Bull 5:11–23. Available at https://www.researchgate.net/publication/348382390_Exploring_Climate_Change_Adaptation_in_Coastal_Communities_of_Bangladesh_Using_Adaptation_Capability_Index_ACI
Islam MF, Rashid AB (2011) Riverbank erosion displaces in Bangladesh: need for institutional response and policy intervention. Bangladesh J Bioethics 2(2):4–19. Available at https://www.banglajol.info/index.php/BIOETHICS/article/view/9540
Islam SMD, Uddin MJ (2015) Impacts, vulnerability and coping with cyclone hazard in coastal region of Bangladesh: a case study on Kalapara Upazila of Patuakhali District. Jahangirnagar Univ Environ Bull 4:11–30. Available at https://www.researchgate.net/publication/345974268_Impacts_Vulnerability_and_Coping_with_Cyclone_Hazard_in_Coastal_Region_of_Bangladesh_A_Case_Study_on_Kalapara_Upazila_of_Patuakhali_District
Kebede YS, Endalamaw NT, Sinshaw BG, Atinkut HB (2020) Modeling soil erosion using RUSLE and GIS at watershed level in the upper beles, Ethiopia. Environmental challenges (Vol. 2, p. 100009). Elsevier BV. https://doi.org/10.1016/j.envc.2020.100009
Keya MK, Harun SMR (2007) Riverbank erosion induced stress and coping of displaced women in Bangladesh. Empowerment 2007(14):17–30
Lein H (2009) The poorest and the most vulnerable? On hazards, livelihoods and labelling of riverine communities in Bangladesh. 30. https://doi.org/10.1111/j.1467 9493.2008.00357.x
León de Villagrán, Carlos J (August 2006). Vulnerability: a conceptual and methodological review. Available at http://www.ehs.unu.edu/file/get/3904
Mallick A, Mallick B (2021) Staying despite riverbank erosion: evidence of coastal Bangladesh. SN Soc Sci 1, 155 (2021). https://doi.org/10.1007/s43545-021-00104-x
MODMR (2021) Trend and impact analysis of internal displacement due to the impacts of disaster and climate change. Ministry of Disaster Management and Relief, Dhaka. Available at https://www.undp.org/bangladesh/publications/trend-and-impact-analysis-internal-displacement-due-impacts-disaster-and-climate-change
Momen MN, Begum MM (2021) Small Island development states (SIDS): nexus of environment and sustainable livelihood. In: Leal Filho W, Azul AM, Brandli L, Lange Salvia A, Wall T (eds) Partnerships for the goals. Encyclopedia of the UN sustainable development goals. Springer, Cham. https://doi.org/10.1007/978-3-319-71067-9_39-1
Momen MN, Ferdous J (2023) Engagement of civil Society in Good Governance and Sustainable Development. In: Governance in Bangladesh. New Frontiers in regional science: Asian perspectives, vol 68. Springer, Singapore. https://doi.org/10.1007/978-981-99-0424-2_6
Moore WH, Shellman SM (2002) Refugee or internally displaced person? To where should one flee? Florida State University, November 22. www.google.com/url?sa¼U&start¼2&q¼http://www.ccis-ucsd.org/PUBLICATIONS/wrkg65.pdf&e¼9888

Munna M, Nayem H (2018) 5 causes and 17 effects of riverbank erosion in Bangladesh. Org, EarthReview. Available at https://www.academia.edu/39573214/5_Causes_and_17_Effects_of_Riverbank_Erosion_in_Bangladesh
Nishat A, Mukherjee N (2013) Climate change impacts, scenario and vulnerability of Bangladesh. In: Shaw R, Mallick F, Islam A (eds) Climate change adaptation actions in Bangladesh. Disaster risk reduction. Springer, Tokyo. https://doi.org/10.1007/978-4-431-54249-0_2
Okeke CAU, Azuh DE, Ogbuagu FU, Kogure T (2020) Assessment of land use impact and seepage erosion contributions to seasonal variations in riverbank stability: The Iju River, SW Nigeria. Groundwater for Sustainable Development. vol. 11, p. 100448). Elsevier BV. https://doi.org/10.1016/j.gsd.2020.100448
Pelling M (2003) The vulnerability of cities. Earthscan Publication Ltd., London, UK 2003: 46–67
Rahman MATMT, Islam S, Rahman SH (2015) Coping with flood and riverbank erosion caused by climate change using livelihood resources: a case study of Bangladesh. Clim Dev 7(2):185–191. https://doi.org/10.1080/17565529.2014.910163
Schneider SH (2007) Climate change: impacts, adaptation and vulnerability. In: Parry ML, Canziani OF, Palutikof JP, Van der Linden PJ, Hanson CE (eds) Contribution of working Group-II to the fourth assessment report of the intergovernmental panel on climate change. Cambridge Univ. Press, Cambridge, pp 779–810
Shamsuddoha M, Chowdhury RK (2007) Climate change impact and disaster vulnerabilities in the coastal areas of Bangladesh. COAST Trust and Equity and Justice Working Group, Dhaka. Available at https://www.preventionweb.net/files/4032_DisasterBD.pdf
Smith K (1992) Environmental hazards: assessing risk and reducing disaster, 2nd edn. Routledge physical environment series. Routledge, London
Thi TD, Minh DD (2019) Riverbank stability assessment under river water level changes and hydraulic erosion. Water 11(12):2598. https://doi.org/10.3390/w11122598
Turner BL, Kasperson RE, Matson PA, McCarthy JJ, Corell RW, Christensen L, et al (2003). A framework for vulnerability analysis in sustainability science. Proceedings of the National Academy of Sciences 2003:100(14):8074–9. https://doi.org/10.1073/pnas.1231335100
Twigg J (2007) Characteristics of a disaster-resilient community: a guidance note. United Kingdom Department for International Development (DFID) disaster risk reduction interagency coordination group. www.benfieldhrc.org/disaster_studies/projects/communitydrrindicators/community_drr_indicators_index.htm
UNDRR (2017) UNDRR terminology, 2017. Vulnerability, Available at https://www.preventionweb.net/understanding-disaster-risk/component-risk/vulnerability
UWICED (2002) Vulnerability and Small Island developing states. University of the West Indies Centre for environment and development, prepared for the UNDP/CAPACITY 21 Programme of action for the sustainable development of Small Island states, Available at http://www.sopac.org.fj/Projects/Evi/Files/Env%20vuln%20of%20SIDS.pdf
Walsham M (2010) Assessing the evidence: environment. Climate Change and Migration in Bangladesh, International Organization for Migration (IOM). Available at https://publications.iom.int/system/files/pdf/environment_climate_change_bangladesh.pdf
Watts MJ, Bohle H-G (1993) The space of vulnerability: the causal structure of hunger and famine. Prog Human Geo 17(1):43–67. https://doi.org/10.1177/030913259301700103

# Chapter 3
# Riverbank Erosion in Bangladesh: Livelihood Risks and Vulnerability Trend and Patterns

**Abstract** A critical concern is riverbank erosion, which has significant economic and livelihood consequences for the impacted populations in Bangladesh. This chapter thoroughly explores the various patterns of risks and vulnerabilities in the livelihoods of IDPs in Bangladesh. The chapter sought to acquire knowledge from previous researchers who contributed to the identification of livelihood vulnerabilities, the exposure of sustainable livelihood adaption approaches, and the risk and vulnerability assessment. Furthermore, the author provided a detailed picture of vulnerability mapping using data obtained from field investigation, following the approaches established by earlier researchers and scholars and the guiding principles of the IPCC. However, this study is founded on the primary data collection in the most river bank erosion-prone areas of Sirajganj district in Bangladesh. This was collected through the carefully selected direct interview of 66 individuals from 54 families, including 20 households from urban slums in Dhaka (railway station slums that are impacted by river bank erosion), FGDs, KIIS and case stories and the detailed justification is provided in the chapter one.

## 3.1 Introduction: Livelihood Risk and Vulnerability Analysis

Bangladesh, a country with topographical and hydrological disadvantages, encounters substantial problems from sea-level rise and associated natural hazards, especially in its coastal areas (Rabbani et al. 2019). Riverbank erosion is a persistent and longstanding phenomenon in Bangladesh, caused by the alteration and diversion of the three principal river systems: the Ganges, Brahmaputra, and Meghna. This problem is not solely attributable to floods; it is a continuous process that profoundly affects the lives and livelihoods of individuals residing along these rivers (Hossain 1993).

Riverbank erosion can have catastrophic economic consequences, resulting in financial distress, agricultural losses, and property damage for impacted populations. The vulnerability to riverbank erosion in Bangladesh is increased by their proximity to mountainous areas in India, reinforcing the risk of flash floods from

 
M. Begum, *Understanding River Bank Erosion in Bangladesh*, Springer Water,
https://doi.org/10.1007/978-3-032-00288-4_3

upstream Indian rivers (Rahman et al. 2018). However, the economy of Bangladesh is heavily reliant on agriculture, and the effects of natural disasters, such as riverbank erosion, pose a substantial impediment to the nation's sustained and steady development. Bangladesh's significant vulnerability to riverbank erosion has created a distinctive living environment, especially in the coastal and northwestern areas, where the marginalized population suffers the most (Rabbani et al. 2019; Bhuiyan et al. 2017; Hossain 1993; Rahman et al. 2018).

Risk analysis is to improve the capacity of decision makers' and those in charge of operations', and vulnerabilities within their own areas of operations, as well as to provide a foundation for their own planning (Rana and Nessa 2017). Moreover, the basis may develop an essential source of information for affected community. The risk and vulnerability analyses of actors also contribute to present a picture of the risks that exist in society as a whole. Thus, risk analyses must accommodate two perspectives: one for their own organization or operation, and the other to meet the need for providing a collective risk profile for society as a whole. This analysis may provide a suitable basis for decisions regarding the objective of the work in the field of crisis preparedness. This may also be used to determine the purpose of practices, training, and the allocation of funds, among other things.

The analyses of risk contribute to reducing the vulnerability of society and enhancing its capacity to manage crises and extraordinary incidents. Analyses of risk provide crucial information for preventing, preparing for, and managing crises to provide a collective risk profile for the entire society (Dewan et al. 2017).

Risk analysis might be explained in this way; Risk = Threat × Vulnerability × Consequence.

UNDRR (2017) declared risk is considered as one of the defining factor of disaster risk, and designed it as; $\text{Risk} = \text{Hazard} \times \text{Exposure} \times \text{Vulnerability}$.

The analysis of risk and vulnerability, both in the World Bank's work and in different situations, focuses on two distinct areas of concern. Initially, it emphasizes the significance of risk in the fluctuation of poverty and the strategies that households employ to mitigate their vulnerability to different types of risks. This analysis examines the influence of risk on poverty, considering both the informal and formal methods households employ to minimize, and manage risk. Within this document, the author use the term "risk and vulnerability analysis" to describe this particular area of research. Furthermore, there is a particular emphasis on specific marginalized populations who possess low capacity for facing poverty and have limited opportunities for breaking free from persistent poverty (Hoogeveen et al. 2004; Lein 2009).

Risk and vulnerability analysis can encompass various aspects of poverty, rather than solely focusing on income, in accordance with the specific demands of the context-specific analysis. Poverty is a concept that goes beyond simply having a poor income or level of consumption (Hoogeveen et al. 2004). However, deprivation is frequently associated with income poverty, such as when a low income hinders someone from attaining enough nourishment or accessing treatments for treatable illnesses. However, poverty is not always associated with income. It can also arise due to limited availability of public services and programs (such as healthcare or education) or the absence of political, civil, and economic freedoms.

Exposure to risk can be regarded as a facet of poverty. Low-income households are generally more vulnerable to risk and have a lower level of protection against it. The connections between poverty and risk establish the concept of vulnerability. Specifically, poverty signifies a distressing level of welfare. However, when poverty is eliminated, vulnerability as the exposure to risk is no longer a concern. While households may still face potential risks, as long as these risks do not significantly impact their overall well-being, the resulting consequences are considered acceptable.

The term *vulnerability* is often used to refer to a state of weakness or defenselessness, particularly when describing groups that are susceptible to significant hardships (Bankoff et al. 2004). These people are at risk of being trapped in long-term poverty, without the necessary resources to benefit from possibilities and having insufficient protection against unexpected events or shocks (Kabir and Kamruzzaman 2022; Hoogeveen et al. 2004). Vulnerability assessments have emerged as a fundamental approach for realizing the impacts of climate change and development challenges in numerous contexts. These assessments may incorporate a wide range of techniques used to methodically examine the physical and social interactions between humans and their surrounding environment (Hahn et al. 2009). It can be said that risk and vulnerability assessment is very essential for sustainable livelihood of people.

## 3.2 Sustainable Livelihoods

River bank erosion is regarded to be a very problematic environmental hazard (Chatterjee and Mistri 2013) that poses a threat to sustainable livelihood. According to Chambers and Conway (1991), sustainable livelihoods can be defined as individuals, and their abilities and resources for survival, which include food, money, and assets. Tangible assets refer to physical resources and stores, such as property or equipment. On the other hand, intangible assets are non-physical and include claims and access to certain benefits or rights. An environmentally sustainable livelihood is one that preserves and improves the national and global resources that livelihoods depend on, while also having positive effects on other livelihoods. A socially sustainable livelihood is one that is capable of effectively coping with stress and unexpected events, while also ensuring the well-being of future generations (Joakim 2012; Elahi and Rogge 1991). Thus, livelihoods encompass the resources and abilities required to support the basic human needs for an extended period of time. Aside from consuming food, shelter, clothes, cultural values, and social relationships are also seen as essential requirements (Gaillard 2015).

The Sustainable Livelihood Approach (SLA) is employed to design community development programs and to comprehend domestic livelihoods (Fig. 3.1). This approach incorporates five distinct categories of assets owned by households-natural, social, financial, physical, and human-and employs a variety of indicators to evaluate the degree of risk and vulnerability to climate change and natural

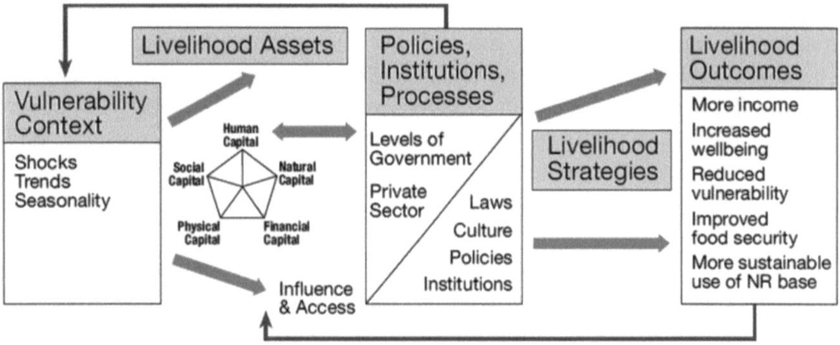


**Fig. 3.1** Sustainable livelihood framework (DFID 2000)

disasters. The economic attributes of households have an influence on their capacity to adapt, while the attributes of health, food, and water resources relate to the vulnerability of households due to the effects of climate change (Chambers and Conway 1991). Figure 3.1 shows the relationships from the vulnerability contexts to sustainable livelihood outcomes. From this point of view multifaceted vulnerability context like seasonal shocking and tends of vulnerability always influence the sustainable livelihood outcome whereas other livelihood assets and sociocultural and other political institutional process act as important variables for better livelihood (Birkmann 2007; Haque and Zaman 1989).

## 3.3 Vulnerability Assessment

Several frameworks, and approaches for vulnerability assessment have been introduced to enhance the theoretical basis and its application (Gallopin 2006). This book considers vulnerability to be determined by both bio-physical risk and social reaction, as well as how these factors contribute to the occurrence of hazards in a particular location (Munna and Nayem 2018; Hewitt and Burton 1971). A semi-structured interview questions were undertaken to assess the level of vulnerability. Data collection was based on the vulnerability issues identified from the interviews, FGD and KIIs sessions. There are two main ways to enhance adaptive capacity, as stated by Huq and Klein (2003). The first approach involves improving the vulnerable systems or processes managed by local communities, allowing for adjustments in organizational and management practices to mitigate the risks of riverbank erosion (Birkmann 2005). The second approach focuses on raising public awareness and preparedness by updating people about the potential consequences of riverbank erosion, including potential adaptation issues or migration concerns (Walsham 2010).

$$\text{Resilience} = \frac{\text{Adaptive capacity}}{\text{Vulnerability}} \quad \text{Source : Mahmood and Mahbub}(2018)$$

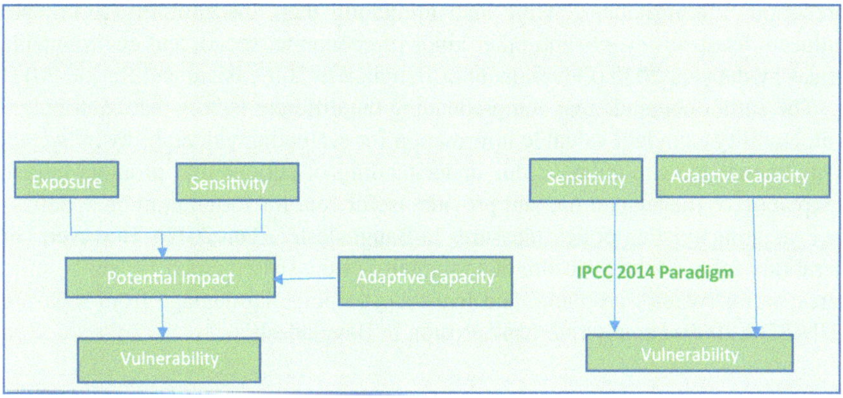


**Fig. 3.2** Vulnerability components (Sharma and Ravindranath 2019)

On the other note, vulnerability assessments continue to account for the exposure component explicitly using the IPCC (2007) definition of vulnerability. IPCC (2014) notes that vulnerability as the inclination or predisposition to experience unfavorable impacts (Sharma and Ravindranath 2019; IPCC 2001).

Although vulnerability is often portrayed as being independent of exposure and risk, planners and decision-makers require information regarding the susceptibility of social and natural systems to climatic threats like landslides, unseasonable rains and riverbank erosion fact. While vulnerability, as indicated by the IPCC (2014) report, is not directly influenced by hazards, it is nonetheless evaluated within the framework of an expected hazard. This is done by selecting indicators that are relevant to the hazard, which measure the sensitivity and adaptive capacity of a system. It makes a major weakness of the adaptation mechanism of a nation (Fig. 3.2).

## 3.4 Various Livelihood Vulnerability Patterns

This section discusses on contributing factors, major components and indicators of vulnerability on exposure, sensitivity, adaptive capacity, based on the suggestion from the IPCC. While identifying human vulnerability in crucial policy frameworks (such as those from the IPCC) and Bhuiya et al. (2017) present a broader viewpoint by investigating multiple components that may led to vulnerability in Bangladesh's river bank erosion prone areas. These indicators include the frequency and intensity of riverbank erosion, and its exposure to disastrous consequences, demographic sensitivity (especially in case of children and senior citizens), absence of agriculture production, presence of critically ill relatives or family members, access to healthcare facilities, prevalence of local illnesses, and inadequate sanitation. These indicators aim to identify livelihood vulnerability within policy frameworks, such as those established by the IPCC (2001). However, target 1.5 of SDG Goal 1 explicitly aims to enhance the resilience capacity of impoverished individuals and those in precarious circumstances, while also mitigating their susceptibility to climate-induced disastrous events and other kinds of economic, social, and environmental crises by the year 2030 (Government of Bangladesh 2009; Islam and Rashid 2011).

The author contends that comprehending the different factors that contribute to vulnerability can yield valuable information for evaluating vulnerability indicators. This knowledge can enhance our understanding of actions that promote climate adaptation or maladaptation, and provide useful data for formulating new policies or evaluating existing policy measures in Bangladesh (Alam 2016). However, vulnerability refers to the exposure of community-critical assets, such as people, structures, infrastructure, cropland, and livestock, to potential damage from a hazard, (UNISDR 2009) namely riverbank erosion in Bangladesh.

## 3.5 The Equation Between Exposure, Sensitivity, and Adaptive Capacity

However, the author attempts to identify patterns of vulnerability based on the three components of *exposure, sensitivity, and adaptive capacity*. High vulnerability is characterized by a combination of high exposure to climate or natural disaster risks, high sensitivity, and low adaptive capacity (Babu 2016). The factors that influence vulnerability are sensitivity, exposure, and adaptive capacity. Low adaptive capacity in comparison to sensitivity and exposure contributes to a high degree of vulnerability (top). On the contrary, increased adaptive capacity mitigates the consequences of greater exposure and sensitivity, thereby diminishing vulnerability (bottom) (see Fig. 3.3).

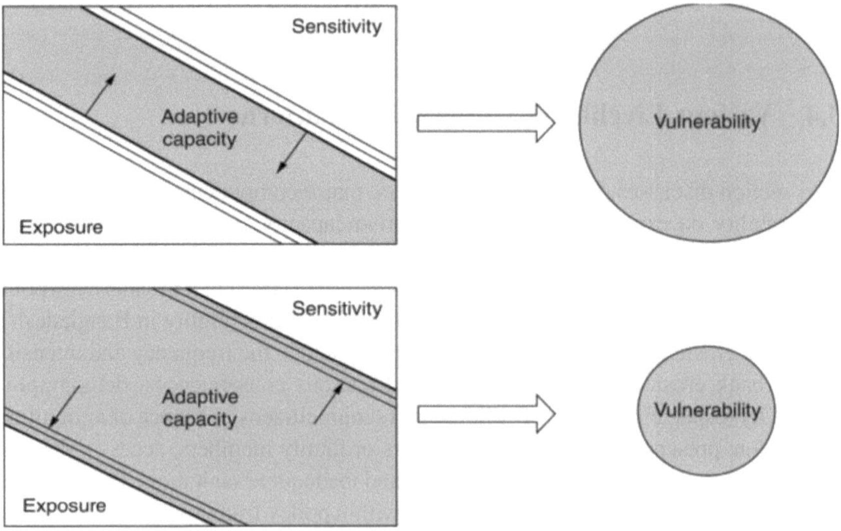


**Fig. 3.3** Equation between exposure, sensitivity, and adaptive capacity (Thomas et al. 2018)

### 3.5.1 Riverbank Erosion (Frequency and Magnitude)

Riverbank erosion, as one of the primary causes of vulnerability, is a significant contributor to poverty at the national level in Bangladesh. Over the recent years, the rate of erosion has accelerated due to climate change, which has heightened the degree of economic vulnerability, such as loss of economic value. Riverbank erosion in Sirajganj district shows varying intensities throughout the year, with the highest likelihood occurring between June and September. However, Sirajganj is identified as the region most susceptible to riverbank erosion in Bangladesh, with a cumulative land erosion rate of 622.2 ha in 2009 (CEGIS 2010; Uddin and Basak 2012).

**Case Story 1**

Md. Rony Seikh, Sharif More, Koizuri, Shahajadpur.

12 Bigha land undergrown near to the Jamuna river, including 10 sotok house-land. At present they are living on khas land.

**Case Story 2**

Md. Shobhan, Upazila Agriculture extension Department, Shahajadpur

Each year every villages lost huge dimension of house land and agricultural land, though the office does not have actual data but it is nearly 1000 Bigha (14,400,000 sq. ft., as 1 Bigha = 14,400 sq. ft.) land lost last 3/4 years.

And the landless people rarely receive their previous land that they lost after char land alive on river, most of the cases the muscles people took the land.

In many cases, when farmer is used to collect their grains, the influential people demand that this is their land.

**Case Story 3**

Md Shajahan [individual], Hatpachil, 2 no. Ward, Koizuri union, according to his voice;

- He has lost his father's household area 5 years before, and if the erosion is continuing in the same way he will lost his own house within next 2 years.
- Due to erosion the amount of relief is really very poor and it cannot meet the demands of the extreme vulnerable people.
- Last 15–20 years this union lost 2 large Mosque, 2 large bazar, 4/5 schools and madrasahs.

### 3.5.2 Human Mortality Rate

As this research is working with the riverbank erosion vulnerability, so the connection of mortality rate is not found here, though few respondents echoed the previous findings that food consumption, level of nutrition and sustainable livelihood can foster the life longevity of this area's population. Although respondents have reported significant number of child death while bathing in the Jamuna River.

### 3.5.3 Percent of HHs Reporting to an Injury

Riverbank erosion is a widespread environmental concern in Bangladesh for its vast system of rivers and the livelihoods depended on them. Located in the Bengal delta, Bangladesh undergoes significant silt deposition from the Ganges, Brahmaputra, and Meghna rivers, resulting in a dynamic and frequently unpredictable terrain. The study of Bhuiyan et al. (2017) seeks to measure the effect of riverbank erosion on the occurrence of home injuries in impacted populations. Riverbank erosion is a persistent and longstanding phenomenon in Bangladesh, especially along the principal river systems (Hossain 1993; Mallick and Mallick 2021). Only a small percentage of community members in the study areas have had physical injuries due to riverbank erosion. The respondents said that they are not physically affected by this hazard; however, their mental state is significantly affected.

### 3.5.4 Total Number of Affected People

Riverbank erosion constitutes a major environmental issue in Bangladesh, especially in the Sirajganj district, where the Ganges, Brahmaputra, and Meghna rivers intersect. The process, propelled by sediment deposition and movement, has resulted in the displacement of many populations and the loss of vital agricultural land (Hossain 1993). The effect of riverbank erosion on the people of Sirajganj district is significant. Estimates indicate that the extent of damage inflicted by this issue remains inadequately acknowledged, both in Bangladesh and internationally. The alteration and diversion of river courses, intensified by floods, have resulted in persistent and prolonged erosion, especially around the three principal rivers and their tributaries (Hossain 1993). Riverbank erosion has profound economic consequences, resulting in substantial financial distress, agricultural losses, and property damage for impacted populations.

Considering the hurdles, the government of Bangladesh and several NGOs have initiated attempts to tackle the issue. The dredging of riverbeds is a prevalent strategies in Bangladesh, undertaken to enhance navigability and procure construction materials. Nevertheless, the enduring effects of these dredging operations on riverbank erosion and the regional environment necessitate additional examination.

Because of riverbank erosion a large number of households are directly or indirectly affected. Having visited the field office, there are no official records available about the exact number of population displacement. From the field level data the researcher tried to get the answer of this particular question but failed to receive the appropriate answer. In a village the people who had the better opportunity to leave the risky zone they shifted and rest of the family are still trying leave and searching better location. Most vulnerable people are who had no option to stay there. From the respondents of Koizuri Union, nearly 13 HHs moved to another places as IDPs.

### *3.5.5 Percentage of Households Reporting Land Degradation and Initiatives During the Past 10 Years*

The Sirajganj district, situated in the north-central region of the country, has undergone substantial land degradation and riverbank erosion, adversely affecting local residents. The predominant cause of land degradation in the region is the extensive practice of brick manufacturing, resulting in topsoil depletion and environmental contamination. The alteration and redirection of significant river flows, including the Padma, Jamuna, and Meghna, have resulted in substantial riverbank erosion, culminating in the loss of arable land and the displacement of inhabitants (Hossain 1993).

**Case Story 4**

Md. Mojjem Hossain (Freedom fighter), Chairman, Koizuri Union

Massive riverbank erosion in this region, on an average, more than 50 families lost their houses each year. Along with this, hundreds of families lost their own agricultural land.

Each year Vulnerable Group Development (VGD) card was given to 261 families which amount is 30 kg rice in each month.

Vulnerable Group Feeding (VGF) was given to 4200 families which amount is 10 kg rice in each Eid, major Muslim festival.

General Relief (GR) was given to each vulnerable family's 2–3 thousand tk and rice.

Affected families given a tube well for safe drinking water in each year.

Distributed the tube wells to various religious institutions such as masjid and madrasa.

GOB has started different projects to manage the riverbank erosion and hopefully we will overcome the situation after finishing the projects.

**Image 3.1** A house on close to of collapse due to river bank erosion. *Location* Omarpur union, Sirajganj District (*Photo credit* The author)

***Percentage of Households experienced displacement in the last 10 years.*** See Image 3.1

**Case Story 5**
Md. Shahin Ali, Farmer, Hatpachil, Koizuri Union, Shajadpur.
In every year, due to riverbank erosion hundreds of houses washed away. During that period from the side of government or Water Development Board taken the actions and through the sand bag to protect erosion. It is as like "treatment to dead people". To him, it would be better to drop sand bags during dry season and it might be fruitful.

## *3.5.6 Average Number of Others Natural Disaster in the Last 10 Years*

Bangladesh is a nation exceedingly susceptible to natural calamities owing to its distinctive geographical position and climate. The country encounters a minimum of one significant disaster annually, resulting in an average economic loss of 3.02%

of its GDP each year (Rahman et al. 2018). Bangladesh, especially in the northwestern parts, faces significant susceptibility to floods, droughts, earthquakes, landslides, and riverbank erosion, resulting in a distinctive living condition (Rahman et al. 2018).

In the last ten years, Bangladesh has faced numerous natural disasters, with riverbank erosion being a persistent and longstanding phenomenon that has resulted in substantial economic loss, agricultural losses, and property destruction (Hossain 1993). The economy of Bangladesh is significantly reliant on agriculture, and a primary obstacle to the sustainable and stable growth of food production is the susceptibility of the population to crop failures and food shortages caused by various natural disasters, which exacerbates poverty (Rahman et al. 2018).

Dhaka, the capital city with a population exceeding ten million, has been identified as one of the 20 most susceptible cities globally to earthquakes, according to an assessment of historical occurrences, exposure, vulnerability, economic factors, transportation infrastructure, emergency response, and recovery capacity (Alam 2019). Over the past decade, Bangladesh has encountered a considerable frequency of natural disasters, as noted by Alam (2019), Rahman et al. (2015), Rahman et al. (2018), and Apu and Das (2020). Riverbank erosion constitutes a significant problem, since the country's extensive delta plain and 230 rivers present an ongoing threat throughout the monsoon season. However, this study area is also the high sensitive area for the risk of thunder storm. One of the respondents said, last year 11 people died due to thunderstorm while they were working in open agricultural land. In other ways, a very few number of people especially child under 12 are missing while flooding situation is worsened.

## 3.6 Health Sensitivity

In this study, health sensitivity is defined as the degree to which the system is affected by the presence of risks and vulnerabilities. This study examines the impacts of riverbank erosion, a natural hazard, on the livelihood patterns of the affected households living in the study area. If individuals are impacted on an annual basis, it can be classified as a high health sensitivity zone. However, if the government is able to effectively mitigate the impacts of the event, it might be classified as having low sensitivity to riverbank erosion in the study area. The criteria listed and explored below are intricately linked to the determination of health sensitivity in the study areas.

### *3.6.1 Average Time to Reach Healthcare Center*

Riverbank erosion is a significant problem in Bangladesh, having extensive implications for the impacted populations. This dynamic process results in the displacement of thousands annually, hindering their access to fundamental services, such as healthcare (Mollah and Ferdaush 2015).

A significant problem encountered by individuals impacted by riverbank erosion is the high distance and time necessary to access healthcare facilities. The erosion of riverbanks may separate communities, necessitating long journeys to obtain medical care. This issue is especially troubling during emergencies or when someone need urgent medical attention, as prolonged travel time can significantly impact their health and well-being. Research has been undertaken in many impacted locations of Bangladesh to understand the effects of riverbank erosion on healthcare access. Research indicates that population displacement caused by riverbank erosion might substantially prolong the time required to access healthcare facilities.

This displacement probably contributed to increased travel times for these persons to obtain healthcare services. A study on the economic impacts of riverbank erosion has underscored the substantial costs incurred by impacted communities, comprising property damage and agricultural losses. These financial pressures might intensify the difficulties in obtaining healthcare, since individuals may find it challenging to afford transportation or encounter other economic obstacles.

The reliable data about average travel times to healthcare facilities during riverbank erosion events is scarce; however, existing research indicates that this is a critical issue demanding greater examination and focused interventions. However, riverbank erosion in Bangladesh can lead to considerable economic consequences incorporating agricultural and property losses, along with the disruption of livelihoods (Hossain 1993).

From the answer of the respondents they need minimum 30 min to reach the Upazila health care center, but the quality of services are not well enough. Most of the cases people have to move the district level hospital for better services though the community clinic are there available to provide primary treatments and suggestions.

### *3.6.2 Percentage of Households Who Have Critical Illness Member*

In the majority of situations, households don't keep the prescription data for future reference. Typically, individuals have limited awareness regarding their annual health check-up. Elderly individuals are present in the research area who are sick, but the other members of their family perceive their health status as a normal consequence of old age. Members of HHs were not interviewed with severe illnesses. Furthermore, the respondents reported that 4 or 5 individuals died in their villages and nearby areas during the COVID-19 pandemic. However, it remains uncertain if these deaths were caused by the coronavirus or other factors.

### 3.6.3 *Percentage of Households in Which at Least One Family Member Has Been Diagnosed with Infectious Disease*

Respondents of the both union are not well informed about communicable diseases. The health issue never got the priority to their livelihoods. From the FGD discussion, the researcher found that better livelihoods means to the river bank areas people that they need safe shelter, security of food, cloths, and education. Health issue is not coming before these. Though few of them thought that health matter is somehow controlled by natural belief, and everything *God Knows.*

The researcher found that riverbank erosion is a significant issue that negatively impacts their mental well-being. The communal people suffer mostly in mental illness like depression, anxiety, stress and trauma. Women households are not exemption on that issues. The women respondents said they are always concern about this fact and even when they are slept. Other respondents said that there is a high prevalence of water-borne diseases, especially among individuals suffering from diarrhea. The researcher found that a mere 4% of the respondents were aware of their experience with communicable diseases in their different areas.

### 3.6.4 *Percentage of Households Lacking Proper Sanitation Facilities and Having Poor Hygiene Conditions*

Figure 3.4, total respondents of (66 people = male 32, female 27, old 7) the researcher found around 46% male, around 40% female and 9% old people uses toilet (different forms) adjacent to their house areas. Beside this, the researcher also found a few number of households who don't have access to toilet facilities. Figure 3.4 shows the data about uses of different forms of toilet, based on gender and age status.

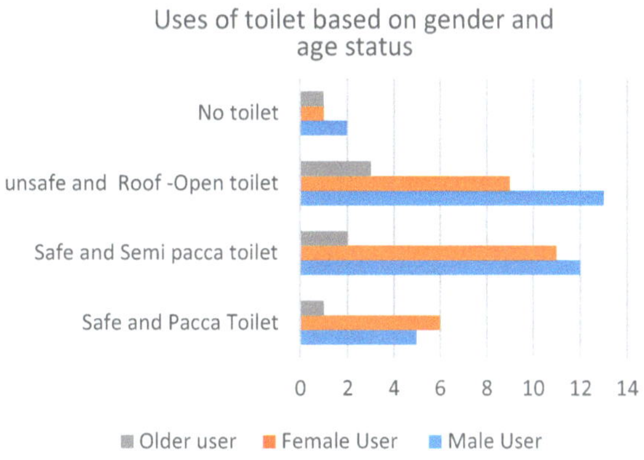


**Fig. 3.4** Uses of toilet based on gender and age status

Public health demands are incurred by the affected people. Those who have been displaced may be forced to survive in the open air, without access to adequate sanitation or potable water systems.

**Case Story 6**

Md. Rashedul Islam, Inspection Officer, Upazila Project Implementation, Shahajadpur

Due to riverbank erosion, he went to visit the area frequently. As a result of riverbank erosion people lost their households and take a shelter to near relatives or friends. As many people live together it looks like a slum and sanitary system become hazardous and unhygienic.

Finally, as the sufferer don't get any support from government side the people are bound to move neighbor villages or other places like nearer city and manage their livelihoods, anyway.

Anil Kumar Gosh, UNO office, Shahajadpur

As a result of riverbank erosion, the Charland is being transformed, allowing the Khas land to expand. Consequently, families who have lost their land can use and cultivate it without paying taxes.

## 3.6.5 *Inverse Life Expectancy*

Bangladesh, located in the Bengal delta, is distinguished by its extensive system of rivers, streams, and canals, altogether about 24,000 km in length (Bhuiyan et al. 2017). This complex river system presents a considerable problem due to riverbank erosion, a persistent and historical process with substantial economic and social consequences for the country (Hossain 1993).

Riverbank erosion is attributed to the alteration or redirection of river channels, intensified by flooding (Hossain 1993). This phenomena significantly affects the lives and livelihoods of those in the affected areas, resulting in economic loses, agricultural losses, and property damage (Hossain 1993). The significance of the issue is emphasized by the estimation that approximately 129,853 individuals are relocated annually owing to riverbank erosion (Mollah and Ferdaush 2015).

The topographical and economic aspects of riverbank erosion in Bangladesh are complex. The silt deposition in the Ganges-Brahmaputra-Meghna basin, the highest globally, facilitates the ongoing alteration and diversion of river systems. This consequently causes riverbank erosion, leading to the loss of property, cultivable land, and infrastructure, affecting the economic and social welfare of the local community (Hossain 1993).

The impact of riverbank erosion on life expectancy in Bangladesh is complex and interconnected. Research indicates that communities impacted by riverbank erosion frequently experience increased vulnerability, forcing displacement in search of economic possibilities elsewhere. This displacement can significantly impair their access to healthcare, education, and other fundamental services, leading to a reduction in life expectancy.

Moreover, the financial effects of riverbank erosion, including the loss of agricultural land and residential properties, can result in decreasing income and increased poverty, both of which are recognized as critical factors influencing life expectancy. The inverse correlation between life expectancy and riverbank erosion in Bangladesh is a complex issue necessitating a thorough and coordinated strategy to tackle the root causes and alleviate the detrimental effects on impacted populations. However, the current life expectancy for Bangladesh in 2024 is 73.82 years, a 0.34% increase from 2023 (https://www.macrotrends.net/global-metrics/countries/BGD/bangladesh/life-expectancy). Life expectancy is increasing, but the quality of life is not increasing in same way, and the sufferings of people make them more helpless to the will to live.

## 3.7 Food and Agricultural Sensitivity

Bangladesh, is susceptible to the adverse impacts of riverbank erosion, an ongoing and significant environmental issue that jeopardizes the livelihoods and food security of its primarily agricultural population (Bhuiyan et al. 2017; Rabbani et al. 2019; Kundu et al. 2020). The relationship between riverbank erosion and its subsequent impacts on food and agricultural systems in Bangladesh is complex and multifaceted. Kundu et al. (2020) indicate that climate change is intensifying natural hazards that cause riverbank erosion, resulting in detrimental effects on agriculture sector in Bangladesh, which relies heavily on natural resources and climatic conditions. Riverbank erosion not only devastates agricultural land but also undermines essential infrastructure, such as irrigation systems and transportation networks, hence jeopardizing farmers' capacity to maintain their livelihoods. The effects of this environmental concern for food security are significant. Bangladesh has a significant population of impoverished people who cannot satisfy their fundamental nutritional requirements, and the detrimental impacts of riverbank erosion on agricultural output further intensify the situation.

The Government of Bangladesh and the concerned research institutions have acknowledged the significance of this issue, as demonstrated by the recent initiatives of the Bangladesh climate change resilience fund and the World Bank to emphasize of climate change adaptation (Dasgupta et al. 2016; Kundu et al. 2020). A diverse approach is necessary to tackle this situation, including both short-term and long-term measures. Given the context, policymakers and scientists must collaborate with agricultural communities to supply essential resources, information, and political advocacy to execute sustainable land use reforms that reconcile the urgent resource demands of the populace with the necessity of climate change adaptation.

The food and agricultural vulnerability resulting from riverbank erosion in Bangladesh is a multifaceted and urgent concern that necessitates a thorough and cooperative approach from policymakers, researchers, and local communities to guarantee the enduring resilience and food security of the country's agrarian population.

In the research areas, riverbank erosion frequently destroys agricultural land and crops, adversely impacting both harvest quality and quantity, hence intensifying food insecurity. Affected individuals are forced to sell their grain stores to evacuate from the riverbank and save further damage to property. Consequently, livelihoods get disrupted due to the loss of income-generating opportunities.

### 3.7.1 *Insufficient Food on Average Throughout the Months*

Hossain (1993) explore, a notable consequence of riverbank erosion is its impact on food security and vulnerability of rural communities. The persistent reduction of cultivable land and the dislocation of communities resulting from riverbank erosion can result in food insecurity, especially during periods of limited food availability. A study on the economic impacts of riverbank erosion in Bangladesh indicates that the extent of damage from this phenomenon is inadequately acknowledged, both domestically and internationally (Hossain 1993). In another study, it presents evidence of the economic loss, agricultural losses, and property damage endured by impacted areas. It was also examined the effects of riverbank erosion on rural households in Bangladesh that the specific source of this erosion displaces around 129,853 individuals each year (Bhuiyan et al. 2017).

This huge displacement may intensify food insecurity, as displaced families endeavor to live new livelihoods and secure food sources. The susceptibility of rural households in Bangladesh to riverbank erosion is exacerbated by the country's vulnerability to natural catastrophes, including floods, droughts, and earthquakes. These hazards can significantly disrupt food production and distribution, resulting in shortages and price variations that disproportionately harm the most vulnerable communities (Bhuiyan et al. 2017; Rahman et al. 2018; Mollah and Ferdaush 2015; Hossain 1993). Further, erosion along the riverbanks hinders agricultural production and reduces the income of vulnerable individuals from their harvests. The deterioration of farmland may increase the susceptibility of those affected. Respondent(s) struggle with the loss of their homes of subsistence, apart from the lack of a place to reside due to river bank erosion.

### 3.7.2 *Inverse Land Productivity*

In the study areas, respondents reported that the extensive use of chemical fertilizers increases the overall production of crops, but it adversely affects the nutritional content and taste of the food. From their perspective, once the farmer starts the use of fertilizer, discontinuing its usage becomes difficult, and the efficacy of bio fertilizer diminishes compared to its previous effectiveness. The farmer believes that if the current scenario continues, there will be a decline in land productivity in future. It is evident that erosion has resulted in the loss of a significant amount of cultivable ground. However, just a small portion of land with highly fertile characteristics may still be found in the char land area.

### *3.7.3 Percentage of Households that Access to Water*

Riverbank erosion is a widespread problem in Bangladesh, resulting in considerable disruption to communities adjacent to the country's primary rivers (Hossain 1993). This erosion is caused by the alteration and diversion of river flows, intensified by floods, leading to significant economic distress, agricultural losses, and property destruction (Hossain 1993). Households in Bangladesh, located along the Ganges, Brahmaputra, have riverbank erosion concern for households. Over fifty percent of participants in a research expressed hesitance to relocate despite ongoing erosion, emphasizing the significance of community support as mitigation measures (Mallick and Mallick 2021). This sentiment underscores the necessity for a comprehensive approach to adaptation planning that takes into account the priorities for livelihood in the river bank areas.

Estimates indicate that the specific cause of riverbank erosion in Bangladesh leads to the displacement of around 129,853 individuals per year. In the Kazipur Upazila of Sirajganj district, 21,961 individuals, representing 9.35% of the total population, have relocated in recent years due to this issue (Mollah and Ferdaush 2015). The findings also highlights the substantial effect of riverbank erosion on Bangladesh's population, with the interruption of water supplies posing a major issue for impacted households. 30% of the respondents in the area stated that they depend on river water for their livelihood, namely for activities like daily washing and bathing. It was found that individuals used water from the river, and the response highlighted the significance of acquiring drinkable water from secure sources like a tube well.

## 3.8 Demographic Sensitivity

Riverbank erosion is a constant and longstanding phenomenon in Bangladesh, especially near the three major rivers and their tributaries. Although partially attributable to floods, it also stems from the alteration or diversion of river courses. The erosion process has resulted in considerable economic distress, agricultural losses, and property destruction, especially for rural households living along the riverbanks (Hossain 1993). A primary element influencing the demographic sensitivity of riverbank erosion is the dependence of Bangladeshi households on agriculture and riverine activities for their sustenance. The degradation of arable land and water resource accessibility caused by erosion might yield catastrophic effects for these populations, compelling them to adapt and seek alternate methods of sustenance.

Furthermore, the examination of riverbank erosion in Bangladesh has indicated that a considerable segment of the impacted people is resistant to relocation, notwithstanding the persistent danger of erosion. This is frequently attributable to the community participation and support systems that these households have cultivated over time. Riverbank erosion has substantial economic consequence, resulting in the loss of property, infrastructure, and agricultural land, hence increasing the vulnerability of impacted populations (Hossain 1993).

The dredging of riverbeds for navigability and construction materials can have enduring effects on the river system and exacerbate erosion in adjacent areas (Islam et al. 2018). The demographic implications of riverbank erosion in Bangladesh underscore the necessity for a holistic and comprehensive strategy to tackle this issue. This may encompass several interventions, including improved flood management measures, sustainable land-use planning, and the establishment of resilient infrastructure and livelihood alternatives for the impacted communities (Bhuiyan et al. 2017; Hossain 1993; Mallick and Mallick 2021; Islam et al. 2018).

### *3.8.1 Dependency Ratio (Percentage of Individuals Below 15 Years and Over 65 Year's Old)*

The researcher found that each family have dependent individuals. The study explored a total of 54 families and found that, on average, there are 3 dependents per family.

### *3.8.2 Percentage of Households Headed by Females*

No homes consisting of a single individual were found that are the head of the family. This study identified seven widows residing in villages, who are currently residing in the residences of their deceased husbands' families.

### *3.8.3 Average Number of Household Members*

This study found that the average number of family members is 5. However, in some instances, not all members reside together due to temporary job commitments.

Table 3.1 Population density (Banglapedia, Bangladesh National Portal 2011)

| Name of Upazilas | Population density |
|---|---|
| Sahajadpur Upazila | Total population 556,350 |
| | Male 280,508 |
| | Female 275,882 |
| | Population density 1481.35 (per sq./km) |
| Chauhali Upazila | Total population 147,143 |
| | Male 179,738 |
| | Female 173,097 |
| | Population density 2221(per sq./km) |

## 3.9 Population Density

The population density of two research areas Shahajadpur and Chauhali Upazila are nearly 1481.35 (per sq./km) and 2221(per sq./km) respectively (Table 3.1).

## 3.10 Adaptive Capacity

Adaptation relates to the systematic development and implementation of strategies aimed at managing the effects of climate change and disasters. This pertains to the evolutionary process through which people becomes well-suited to its surroundings (Haque 1987). Put it simply, adaptation responds to climate change and environmental stimuli that impacts on people or systems, as defined by the IPCC (2001). Adaptations apply to the internal processes and modifications implemented by human systems to respond to current or future climatic stressors or its effects (O'Brien et al. 2004).

### *3.10.1 Economic Capacity*

*Percentage of households depending exclusively on agriculture.*

**Case Story 7**
Sahera Khatun, a mother of seven children voiced her concerns [location: Monkosha, Sharikmore, Koizuri Union]

Their agricultural land, measuring 4/5 bigha, and residential areas spanning 8–10 sotok, were entirely lost as a result of river bank erosion. Her eldest son is a student at Tongi College, and they are right now unable to cover the household expenditures.

***The average number of households that bear the burden of loan.***

**Case Story 8**

Hafez Abdul Hannan, Imam, Jalalpur Baitulfalah Jame Masjid, Shahjadpur

Currently residing in Jalalpur, but previously lived in Chauhali village. Currently serving as the Imam of a Masjid and as a teacher at a Madrasha.

In early 2019, the entire community collaborated in an effort to mitigate riverbank erosion by constructing a bamboo fence. Approximately 1.2–1.5 million BDT were spent on this project, which was contributed by the local community. The initiative was highly successful, but it was halted when the government assumed the duties associated with it.

***Percentage of households with access to financial services.*** Around 73% of Households receive loans from various NGOs and financial institutions with a high interest rate.

***Percentage of households works outside their own village.***

Only a mere 7% of the overall respondents are involved at distant places in other jobs for their daily needs.

***Household income per month.*** Households need animal feed to prevent further animal fatalities resulting from the flooding of fields and animal feed, considering that their livestock is crucial for their livelihood. Most internally displaced persons (IDPs) are employed as manual laborers or rickshaw pullers. However, the collapse of roads may make rickshaws and other forms of transportation no longer usable. A substantial portion of the impacted community remains jobless due to a scarcity of employment opportunities.

## *3.10.2 Social Network and Communications*

***Percentage of households received assistance from their neighbors within the last month.*** The majority of respondents answered 'no', with only two household members indicating that they have a relative living in Dhaka city who would be willing to assist them in finding employment. Percentage of households who have provided support and assistance to their neighbors over the last month. The researcher found just three individuals who support their neighbor's purchase of land for their future home.

***Displaced who seek protection are susceptible to exploitation, abuse, and violation.*** This fact and occurrence was uncovered after the researcher gathered data from vulnerable individuals in Dhaka city. Respondents in slum areas and new places express a significant level of anxiety for the well-being of their daughters. Additionally, female-headed families are occasionally declined employment opportunities in alternative locations.

***Percentage of households associated with any organization.*** This answer is not received appropriately, all households are involved in both formal and informal organizations to sustain their livelihoods. Average time required to reach the next vehicle station. It is anticipated that the existence of damaged roads will hinder the ability to move of affected individuals and increase the duration of travel to the affected areas. The dispersion of the affected population may present difficulties in identifying and accessing victims. There is a lack of information regarding any further limitations on access.

***Percentage of households that possess communicative electronic devices such as televisions and cell phones***. 90% of the people possess cell phones as means of communication. The respondents expressed their current preference for mobile phones, despite the fact that just 58% of them own a television.

### *3.10.3 Education and Skills*

***Education index***. Within the study areas, there exist four distinct categories of academic institutions. The educational institutions mentioned are: primary school, High school, College, and Madrasah. The education of both children and youth is severely impacted by the critical period of riverbank erosion in. Furthermore, local educational institutions occasionally experience discontinuation due to erosion. During that period, the school activities were relocated to a different location, resulting in the withdrawal of female students. Through discussions with teachers from schools, colleges, and madrasahs, this research has found a significant disparity in the number of students enrolled at the entry level and completion level. Between 15 and 28% of girls in the study area drop out of school at the beginning of secondary school, and one of the main factors contributing to this is when families relocate to new places. Once again, a significant proportion of adolescent girls, ranging from 30 to 40%, discontinue their education as a result of early marriage and domestic responsibilities. Male children also experience similar hardships. 10% of school children and 15% of secondary boys drop out from their previous school. This study revealed that madrasah students have a lower dropout rate despite the need to transfer between institutions. According to this research, the study area revealed that a mere 9% of students discontinue their education from their previous institution.

***Percentage of homes with individuals possessing any type of formal or informal skills***. According to the villagers, they possess a diverse range of skills, primarily centered on informal activities such as agricultural labor. Additionally, some of them expressed that they are capable of performing any task given the appropriate instruction.

**Case Story 9**

Md Nurul Islam, previous land Sharif More, Koizuri, Shahajadpur (Now lives at Ullahpara)

He has lost 4.5 bigha of land that he owned.

His father's landholding decreased by 18 bighas.

The area of the house's land is 64 sotok, and the current residence is in a different Thana/Upazila.

He leads a precarious life with four children and regularly visits his char-land field every 7 to 10 days with hope.

**Case Story 10**

Shukur Ali Bepari, village Hatpachil, Koizuri, Shahajadpur

The initial area of his house was 32 sotok, but now only 10 sotok remains due to erosion. Ha once owned 16 bigha of agricultural land, however it has recently been eroded. Due to poverty, he had to withdraw his three female daughters out of five from their education. Previously, he had a modest income. Both he and his wife are currently employed as day laborers. They believed that Jamuna had taken everything from their lives. They made a financial contribution of 30 thousand taka towards a government project aimed at mitigating riverbank erosion. However, they have not received a full refund as the project has not been executed. The individual received a refund of 20 thousand taka.

## 3.11 Conclusion

Riverbank erosion is a natural system that has an adverse impact on livelihood risks and vulnerabilities. No doubt, as Bangladesh suffers from severe river bank erosion and the effects of river bank erosion on human and society are multifaceted. Historically, the environmental issues are the primary cause of people's displacement, although population displacement in Bangladesh due to climate reasons are not an unusual phenomenon. This chapter extensively examines the different patterns of risk and vulnerability in livelihoods in areas affected by and at risk of riverbank erosion in Bangladesh. Furthermore, this chapter provides a full description of the components of vulnerability and their specific indicator, with a particular focus on the study area. However, it is understood that the mapping or assessment of vulnerability is contingent upon the adaption mechanism of specific risks within a well-functioning adaptive system. This chapter analyzes the many patterns of risk and vulnerability experienced by internally displaced persons (IDPs) and their current and future impact on local livelihood in Bangladesh.

## References

Alam GMM (2016) An assessment of the livelihood vulnerability of the riverbank erosion hazard and its impact on food security for rural households in Bangladesh. PhD Thesis, School of Commerce, University of Southern Queensland, Australia, 2016. Available at https://research.usq.edu.au/item/q4qyq/an-assessment-of-the-livelihood-vulnerability-of-the-riverbank-erosion-hazard-and-its-impact-on-food-security-for-rural-households-in-bangladesh

Alam E (2019) Importance of long-term earthquake, tsunami and tropical cyclone data for disaster risk reduction in Bangladesh. Progr Disaster Sci 2, 100019. Elsevier BV. https://doi.org/10.1016/j.pdisas.2019.100019

Apu N, Das U (2020) Tectonics and earthquake potential of Bangladesh: a review (Review of tectonics and earthquake potential of Bangladesh: a review). Int J Disaster Resilience Built Environ 12(3):295–307. Emerald Publishing Limited. https://doi.org/10.1108/ijdrbe-06-2020-0060

Babu S (2016) Exposure, sensitivity and adaptive capacity: understanding climate change vulnerability. Eco-intelligent™. Available at-https://eco-intelligent.com/2019/10/22/exposure-sensitivity-and-adaptive-capacity/

Bankoff G, Frerks G, Hilhorst D (eds) (2004) Mapping vulnerability, disasters, development and people. Earthscan Publishers, London

Bhuiyan MAH, Islam SMD-U, Azam G (2017) Exploring impacts and livelihood vulnerability of riverbank erosion hazard among rural household along the river Padma of Bangladesh. Environ Syst Res 6(25):1–15. Springer Science+Business Media. https://doi.org/10.1186/s40068-017-0102-9

Birkmann J (2005) Danger need not spell disaster—but how vulnerable are we? Number I, United Nations University, Tokyo, Research Brief

Birkmann J (2007) Risk and vulnerability indicators at different scales: applicability, usefulness and policy implications. Environ Hazards 7(2007):20–31. https://doi.org/10.1016/j.envhaz.2007.04.002. Available at www.elsevier.com/locate/hazards

CEGIS (2010) Prediction of riverbank erosion along the Jamuna, Ganges and Padma rivers in 2010. North South University Library OPAC. Available at https://opac.northsouth.edu/cgi-bin/koha/opac-detail.pl?biblionumber=23672&shelfbrowse_itemnumber=37605#shelfbrowser

Chambers R, Conway G (1991) Sustainable rural livelihoods: practical concepts for the 21st century. Discussion Paper 296, Brighton: IDS. Available at https://www.ids.ac.uk/download.php?file=files/Dp296.pdf

Chatterjee S, Mistri B (2013) Impact of river bank erosion on human life: a case study in Shantipur Block, Nadia District, West Bengal. Int J Humanit Soc Sci Invention 2(8):108–111. Available at https://ijhssi.org/papers/v2(8)/Version-3/Q028301080111.pdf

Dasgupta S, Huq M, Wheeler D (2016) Drinking water salinity and infant mortality in Coastal Bangladesh. Water Econ Policy 2(1):1650003. World Scientific. https://doi.org/10.1142/s2382624x1650003x

Dewan A, Corner R, Saleem A, Rahman MM, Haider MR, Rahman MM, Sarker MH (2017) Assessing channel changes of the Ganges–Padma River system in Bangladesh using landsat and hydrological data. Geomorphology 276:257–279. https://doi.org/10.1016/j.geomorph.2016.10.017

DFID (2000) Department for International Development. Achieving sustainability: poverty elimination and the environment

Elahi KM, Rogge RJ (1991) Riverbank erosion, flood and population, displacements in Bangladesh: In: Elahi KM, Ahmed KS, Mafizuddin M (eds) Riverbank erosion, flood hazard and population displacement in Bangladesh, Dhaka. Riverbank Erosion Impact Study (REIS), p 364.

Gaillard JC (2015) Vulnerability and livelihoods. In: People's response to disasters in the Philippines. Disaster studies. Palgrave Macmillan, New York. https://doi.org/10.1057/9781137484291_4

Gallopin GC (2006) Linkages between vulnerability, resilience, and adaptive capacity. Glob Environ Chang 16:293–303. https://doi.org/10.1016/j.gloenvcha.2006.02.004

Goswami S, Rahman SA, Alam MMT, Rahman M, Rafiq MR, Jaman MH, Roy DK (2022) Assessment of shoreline changes and the groundwater quality along the coast of Kuakata, Patuakhali, Bangladesh. J Ecol Eng 23(7):323–332. https://doi.org/10.12911/22998993/149938

Government of Bangladesh (2009) Bangladesh Climate Change Strategy and Action Plan (BCCSAP). Ministry of Environment and Forest. Dhaka: September 2009. Available at https://policy.asiapacificenergy.org/sites/default/files/Bangladesh%20Climate%20Change%20Strategy%20and%20Action%20Plan%202009.pdf

Hahn MB, Riederer AM, Foster SO (2009) The livelihood vulnerability index: a pragmatic approach to assessing risks from climate variability and change-a case study in Mozambique. Glob Environ Change 2009(19):74–88. https://doi.org/10.1016/j.gloenvcha.2008.11.002

Haque C.E. (1987). Impact of riverbank erosion in Kazipur: an application of landsat imagery. REIS Newsletters: Jahangirnagar University, Savar, Dhaka, Bangladesh

Haque CE, Zaman MQ (1989) Coping with riverbank erosion and displacement in Bangladesh: Survival strategies and adjustments. Disasters 13 (4):300–314. https://doi.org/10.1111/j.1467-7717.1989.tb00724.x

Hewitt K, Burton I (1971) The hazardousness of a place: a regional ecology of damaging events. University of Toronto Press, Toronto

Hoogeveen J, Tesliuc E, Vakis R, Dercon S (2004) A guide to the analysis of risk, vulnerability and vulnerable groups. The World Bank, Washington, DC, USA

Hossain M (1993) Economic effects of riverbank erosion: some evidence from Bangladesh. Disasters 17(25–32). Wiley. https://doi.org/10.1111/j.1467-7717.1993.tb00485.x

Huq S, Klein RJ (2003) Adaptation to climate change: why and how. Available at https://www.scidev.net/global/policy-brief/adapting-to-climate-change-why-and-how.html

IPCC (2001) Impacts, adaptation and vulnerability. Summary for policymakers. Intergovernmental panel on climate change (Working group II) 2001. Available at https://www.ipcc.ch/site/assets/uploads/2018/03/WGII_TAR_full_report-2.pdf

IPCC (2007) Climate change 2007: the physical science basis: contribution of working group I to the fourth assessment report of the intergovernmental panel on climate change. Cambridge University Press, Cambridge; New York. Available at https://www.ipcc.ch/site/assets/uploads/2018/02/ar4-wg1-frontmatter-1.pdf

IPCC (2014) Climate change 2014: impacts, adaptation and vulnerability, Part A: global and sectoral aspects, contribution of working group II to the fifth assessment report of the intergovernmental panel on climate change. Cambridge University Press, Cambridge. Available at https://www.ipcc.ch/site/assets/uploads/2018/02/WGIIAR5-FrontMatterA_FINAL.pdf

Islam MK, Kibriya NA, Dustegir MM (2018) Impact analysis of sand dredging from alluvial tidal river. E3S Web of conferences, vol 40, p 03036). EDP Sciences. https://doi.org/10.1051/e3sconf/20184003036

Islam MF, Rashid AB (2011) Riverbank erosion displaces in Bangladesh: need for institutional response and policy intervention. Bangladesh J Bioethics 2(2):4–19. Available at https://www.banglajol.info/index.php/BIOETHICS/article/view/9540

Joakim E (2012) Reducing vulnerability and building resilience in the post-disaster context: a case study of the 2006 Yogyakarta Earthquake recovery effort. Jurnal Sains Teknologi Lingkungan 4. 01-14. https://doi.org/10.20885/jstl.vol4.iss1.art1.

Kabir ME, Kamruzzaman P (2022) Exploring the drivers of vulnerability among disadvantaged internal migrants in riverbank erosion prone areas in North-West Bangladesh. J South Asian Dev 17(1):57–83. https://doi.org/10.1177/09731741221078741

Kundu S, Kabir ME, Morgan EA, Davey P, Hossain M (2020) Building coastal agricultural resilience in Bangladesh: a systematic review of progress gaps and implications. Climate 8(9):98. https://doi.org/10.3390/cli8090098

Lein H (2009) The poorest and the most vulnerable? On hazards, livelihoods and labelling of riverine communities in Bangladesh. Singapore J Trop Geogr 30(1):98–113. https://doi.org/10.1111/j.1467-9493.2008.00357.x

Mahmood R, Mahbub AQM (2018) Building vulnerable islander resilience to natural hazard: a participatory approach. In: Leal Filho W (ed) Handbook of climate change resilience. Springer International Publishing AG, part of Springer Nature 2018. https://doi.org/10.1007/978-3-319-71025-9_107-1

Mallick A, Mallick B (2021) Staying despite riverbank erosion: evidence of coastal Bangladesh. SN Soc Sci 1:155. (2021). https://doi.org/10.1007/s43545-021-00104-x

Mollah TH, Ferdaush J (2015) Riverbank erosion, population migration and rural vulnerability in Bangladesh (A case study on Kazipur Upazila at Sirajgonj District). Environ Ecol Res 3:125–131. https://doi.org/10.13189/eer.2015.030502

Munna M, Nayem H (2018) 5 causes and 17 effects of riverbank erosion in Bangladesh. Earth Rev Org. Available at https://www.academia.edu/39573214/5_Causes_and_17_Effects_of_Riverbank_Erosion_in_Bangladesh

O'Brien K, Sygna L, Haugen JE (2004) Vulnerable or resilient? A multi-scale assessment of climate impacts and vulnerability in Norway. Climatic Change 64:193–225. https://doi.org/10.1023/B:CLIM.0000024668.70143.80

Rabbani G, Munira S, Saif S (2019) Coastal community adaptation to climate change-induced salinity intrusion in Bangladesh. IntechOpen. https://doi.org/10.5772/intechopen.80418

Rahman N, Ansary MA, Islam I (2015) GIS based mapping of vulnerability to earthquake and fire hazard in Dhaka city, Bangladesh. Int J Disaster Risk Reduction 13(291–300). Elsevier BV. https://doi.org/10.1016/j.ijdrr.2015.07.003

Rahman M, Hiya H, Auyon S, Islam M (2018) Exploring the status of disaster risk reduction focusing coping strategies in Rangpur division of Bangladesh. Progressive Agric 29(3):195–204. https://doi.org/10.3329/pa.v29i3.40004

Rana MS, Nessa AM (2017) Impact of riverbank erosion on population migration and resettlement of Bangladesh. Sci J Appl Math Stat 5(2):60–69. https://doi.org/10.11648/j.sjams.20170502.11

Sharma J, Ravindranath N (2019) Applying IPCC 2014 framework for hazard specific vulnerability assessment under climate change. Environ Res Commun 1. https://doi.org/10.1088/2515-7620/ab24ed

Thomas K, Hardy D, Lazrus H, Mendez M, Orlove B, Rivera-Collazo I, Roberts J, Rockman M, Warner B, Winthrop R (2018) Explaining differential vulnerability to climate change: a social science review. Wiley Interdisc Rev Clim Change 10:e565. https://doi.org/10.1002/wcc.565

Uddin AFMA, Basak JK (2012) Effects of riverbank erosion on livelihood. Unnayan Onneshan Innov Dhaka, Bangladesh. https://doi.org/10.13140/RG.2.2.24933.37609/1

UNDRR (2017) UNDRR Terminology, 2017. Vulnerability. Available at https://www.preventionweb.net/understanding-disaster-risk/component-risk/vulnerability

Walsham M (2010) Assessing the evidence: environment, climate change and migration in Bangladesh. International Organization for Migration (IOM). Available at https://publications.iom.int/system/files/pdf/environment_climate_change_bangladesh.pdf

# Chapter 4
# Riverbank Erosion in Bangladesh: Impacts on Physical and Societal Vulnerabilities

**Abstract** Bangladesh suffers from a lack of comprehensive national data, precise figures categorized by villages, union or sub-district, and trends regarding people's displacement. However, the primary data collection in the most river bank erosion-prone areas of Sirajganj district in Bangladesh is the foundation of this study. This was carried out by conducting a meticulously selected direct interview with 66 individuals from 54 families, including 20 households from urban slums in Dhaka (railway station slums that are affected by river bank erosion), FGDs, KIIS, and case studies. The detailed justification is provided in chapter one. However, this chapter examines the extent to which various issues of risk and vulnerability affect socio-economically underprivileged IDPs due to riverbank erosion-prone in Bangladesh.

## 4.1 Introduction: Riverbank Erosion in Bangladesh

The majority of the country's main rivers are Transboundary Rivers that natural flow of water into Bangladesh from India (Uddin and Rahman 2011). Since prehistoric times, riverine Bangladesh has accommodated water flows from the Himalaya and Indo-Burman Ranges that carrying a great deal of sediments as a blessing to people's livelihood and agricultural land. In contrast, excessive water flow through a river channel causes river bank erosion, which poses a threat to livelihood (Das and Bela 2011). This phenomenon erodes the bank directly, relocates people's livelihood, and degrades agricultural land. However, climate and tectonics of the river flow as well as the volume of sediment passing through the channel govern the river's dynamics (Mollah and Ferdaush 2015).

Based on data from the Department of Disaster Management (DDM), riverbank erosion, a consequence of ongoing channel shifting, displaces an estimated 500,000

 
M. Begum, *Understanding River Bank Erosion in Bangladesh*, Springer Water,
https://doi.org/10.1007/978-3-032-00288-4_4

individuals annually along the three main rivers, such as Jamuna, Padma, and Meghna in Bangladesh (Baqee 1997).

River bank erosion happens year-round and due to this disaster, a significant number of people become homeless, thousands of affected households in these areas are uprooted and forced to relocate domestically every year. By erosion, rivers tend to be in a state of equilibrium with respect to their magnitude and intensity. Islam and Rashid (2011) note that erosion occurs when the water overflow and sediments exceeds the resistance capacity of the bank. The erosion of one riverbank results in the deposition of another riverbank. All of these are associated with riverbank erosion and flooding (Elahi et al. 1991). Hasan et al. (2018) observe that along the major rivers of Bangladesh, erosion processes vary and can also be influenced by human activities.

In the 1970s and 1980s, vulnerability was frequently associated with physical structure and its fragility (Bankoff et al. 2004). Vulnerability today encompasses much more than the likelihood of physical structures collapsing. Measuring vulnerability incorporates both quantitative and qualitative methods for describing and operationalizing the phenomena (Birkmann and Wisner 2006). Vulnerability as situations is determined by physical, social, economic, and environmental systems or processes that develop society's exposure to the effects of disaster (UN/ISDR 2004). The impacts of climate change on the longitudinal effects of river bank erosion are currently unknown, it is likely that increased monsoon rainfall could exacerbate the existing situation (Walsham 2010; Rabbi et al. 2013). Since the early 1990s, river erosion rates have been decreasing, for instance, with erosion along the Jamuna river from a high of 5000 hectares of riverbed in every year in the 1980s to approximately 2000 hectares of riverbed in every year since 2000s (CEGIS 2010).

According to BCCSAP (2009), despite this, the rates continue to be substantial enough to disrupt the livelihoods of thousands (as evidenced by the 2010 projections). Furthermore, although the exact consequences are uncertain, there is assumption that an increase in monsoon precipitation could add to heightened erosion tolls along the major rivers of Bangladesh, potentially leading to more extensive long-term damage to residential properties and agricultural land (Haque 1988).

It is pointed out that Bangladesh has faced a significant challenge posed by the growing impact of such natural disasters on agricultural production, though few researchers also argue short-term, such as, agricultural production, income generation, and resources and long-term direct and indirect impacts, such as lives of affected people, children's education and the health status of specific groups (i.e. mothers and children) (Bardsley and Hugo 2010). It is particularly concerned about those who have been relocated as a result of erosion's negative effects (Islam et al. 2013; Islam 2018).

Displacement has long been recognized in Bangladesh and there have been a limited number of research (Hutton and Haque 2003) about the impacts of loss of land on affected people (Walsham 2010). The earlier studies highly focused on displaced families and the potential for disputes over new or reemerging land after flash flood is over (Kabir and Kamruzzaman 2022). However, it is important to reveal that affected households have varying adaptation strategies to cope with it (Dekaraja and Mahanta 2021).

While some displaced people choose to live in the city for longer period of time, others may decide to do so for shorter periods of time (for instance, during tough economic times) in order to help relatives who are still living in river bank erosion prone areas (Rees et al. 2016). However, Hunter et al. (2015) contends that while various hazards have varying degrees of impact on different groups of people, short-term and long-distance migrants' vulnerability patterns can be affected by the features of natural disasters (Ogie and Pradhan 2019).

Due to large population densities, South Asian countries exhibit particularly riverbank erosion and displacement problems (Haque and Zaman 1989). River bank erosion in Bangladesh takes place from May to June, however due to climate change, erosion now occurs from June to August. In Bangladesh, research on riverbank erosion has mostly emphasized on geographical and geological changes happened in case of land formation and river watercourse during the process (Dewan et al. 2017; Kabir and Kamruzzaman 2022). In similar settings, various marginalized groups, especially the economically disadvantaged, or racial minorities, are classified as the most impoverished individuals (Lein 2009). Although there are debates due to over simplification for understanding risks and vulnerabilities along the rivers in Bangladesh (Lein 2009), because many affluent households become landless once affected due to riverbank erosion (Kabir and Kamruzzaman 2022).

## 4.2 Mapping the River Bank Erosion Risks and Consequences

IDPs/EDPs in this study areas, households were compelled to flee their ancestral homes and relocate to adjacent rural areas or within a few kilometers of their original residences, although a group of respondents relocated fifteen to twenty kilometers away from their initial locations. In the earlier research, the process of relocating is a multidimensional decision that is influenced by various factors (Zaman and West 1991), and these include the availability of lands for resettlement, the quality of livelihood options, the accessibility of khas lands (lands owned by the government), the patronage of powerful village leaders, financial assistance from family and friends, and the maintenance of ties with Samaj (traditional social coalitions). Additionally, several other determinants may also play a role (Rashid 2013). People and their way of life are negatively impacted by river bank erosion, as their financial, societal, and mental stress increases over time (Abrar and Azad 2007). Especially, the majority of IDPs in Dhaka is vulnerable and their economic situation is precarious (Fig. 4.1).

The objectives of developing instruments for assessing the extent of vulnerability is to assist in bridging the gap between theoretical foundation and practical aspects of vulnerability (Black et al. 2011) Vulnerability should be viewed as a process

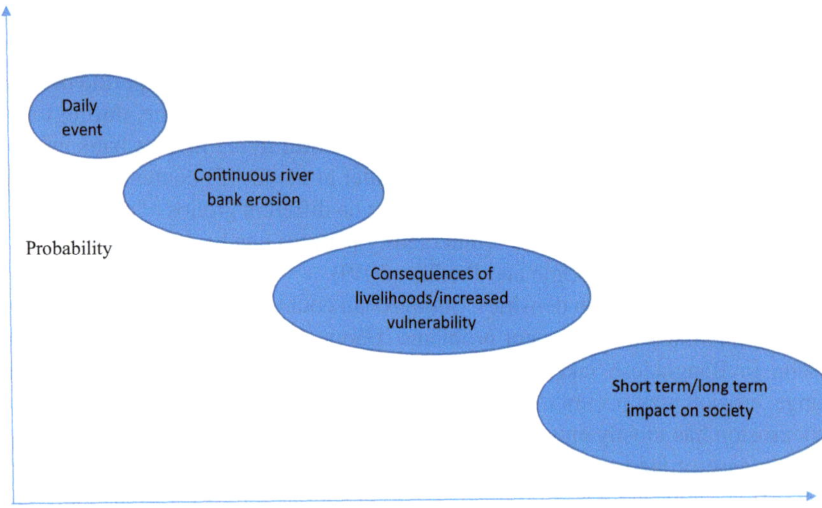


**Fig. 4.1** Vulnerability mapping (the author)

which defines measures and instruments that evaluate the historical, present, and prospective risky or vulnerable areas and populations. In addition to generating improved data for risk assessment, it is essential to reinforce global and local cooperation.

## 4.3 Forced Displacement in Bangladesh

Forced displacement has emerged as a critical issue in Bangladesh, with riverbank erosion serving as a major contributing factor. Riverbank erosion, a persistent and enduring phenomenon in the country, is mainly attributed to the alteration or diversion of river courses, intensified by flooding (Hossain 1993). The geographical and economic aspects of this phenomena have significantly affected the lives and livelihoods of Bangladeshi people. The economic consequences of riverbank erosion in Bangladesh are significant, resulting in financial distress, agricultural losses, and property damage for impacted populations.

Ferris (2011) defines forced displacement that denotes to the IDPs and people displaced, due to natural or artificial disasters, such as chemical related disasters, economic depression, or new development projects. Riverbank erosion hurts local livelihoods. For example, a typical scenario is that the legitimate owner is prevented from recovering his lost property. Vulnerable people afraid due to lack of an effective administrative surveillance mechanism by the government, as the new char land reemerged along the river, it was quickly occupied by the powerful group and the people's vulnerabilities goes higher. Further, the study area in Bangladesh relies on agriculture and river related occupation and the riverbank erosion makes farming unsuccessful. This research found that the respondents opined that the vulnerable people are typically moved semi-permanently or permanently when compelled to avoid the experience again. However, it took time to decide to relocate, even when erosion was imminent. The following Table 4.1 shows the data if the riverbank grabs their homeland what will they do as their perception.

Displacement relates to the vulnerable people of the affected areas and sometimes it influence the push indicators because of strong sociological network (Lein 2009). People with vulnerability intend to avoid it if they get any chance to avoid it (WB 2013). Displacement is the process in which the community feel their ultimate destination, it may be any other zone that is less risky that the original one. Beyond this, people have less trust, expecting that the local authority will support them, although based on their previous experience the majority of the local population were forced to leave their origin of home.

In the majority of cases, under this study, adaptation or mitigation and preparation processes are insufficient to assist the livelihoods. It is perceived that building a resilience capacity through adaptation process at the local level is a priority agenda for affected individuals; when this strategy fails, individuals relocate. Walsham (2010) notes that it is crucial to adopt a positive strategy for subsistence, thereby reducing risks and vulnerabilities. Das et al. (2014) also examined the effects of the Ganges river region of West Bengal in India and the study reveals that riverbank erosion has a variety of social, economic, health, educational, and sometimes political effects, which can lead to displacement.

As stated earlier, a very few research was conducted to determine the nature of displacement resulting from river bank erosion (Siddiqui 2010). Migration is a way for individuals to cope up with the risks associated with vulnerability, however, the definitive relationships between migration and vulnerability are challenging to establish (Laczko and Aghazarm 2009). Migration is typically such as a move to less affected areas (Hunter 2005); although, this may not be easy if the surrounding area for migration is already densely populated (Adamo 2008; Mahmood 1986; Siddiqui 2010). Furthermore, in the study areas, it appears that people has not shown interest in permanent migration due to uncertainty for their future livelihood.

**Table 4.1** Shelter patterns of vulnerable households (the author)

| Groupings (N = 66) | No. of households |
|---|---|
| Personal lands | 8 |
| Purchase lands | 9 |
| Go through relatives house | 7 |
| On the land of others (rent-based) | 17 |
| On khas lands (rent-based) | 12 |
| Adjacent town (Sirajganj) and Dhaka city | 13 |
| Total | 66 |

It is also important to make a difference between mobility, which refers to movement within the specific boundaries, and migration, which occurs when individuals go beyond their boundaries. This division is based on geographical boundaries. When a displaced person moves to a city, she/he faces a mobility challenge because of the new location, situation and ecosystem as well. This movement necessitate a new way of life within new community. For instance, the migrant has been socialized within new settings, she/he must now acquire the urban social structure after moving from a different location. Most definitions of migration focus solely on the physical movement from one location to another.

Migration, in many cases, can be found a manifestation of risks and vulnerability of livelihood, but it can also be a rational strategy for diversifying livelihoods and coping with changes (Walsham 2010). However, adopting policy for displaced people can decrease risks and vulnerability to livelihoods, and ecosystems that may support to the generation and diversification of income, thus improving the capacity of households to cope with the negative effects of resultant changes in climatic conditions.

Due to river bank erosion, forced migrants encounter a variety of insecurities, for example, economic risks and vulnerability due to unemployment, and lack of assets; social insecurity because of the denial of civic and political rights; and health insecurity because of the lack of access to healthcare, among others. However, the living conditions of the displaced people are underprivileged, and try to adapt their lives and livelihood through the low wages, such as rickshaw pulling and daily labor. Das et al. (2014) argue that there should be definitive guidelines to safeguard these rights of displaced people. Apart from the other effects, the damages of agriculture land, livestock, and homes made families exposed to poverty (Chatterjee and Mistri 2013).

During the displacement process, friends and family assisted to some of the households. This research found that fifteen displaced individuals migrated to a location where they had family or acquaintances. It is observed that some households from the study area migrated to Sirajganj district and were doing well financially and they encouraged to other affected individuals to migrate in urban area. These respondents did not receive any assistance from the government or any other bodies in finding employment or shelter, or in acquiring information for rehabilitation. Image 4.1 shows a vulnerable woman and her son observing the damage caused by riverbank erosion, as their home is in close proximity to the river.

## 4.4 Nexus Between Displaced People and Human Security

Human security concerning displaced people reveals a wide range of issues, and this needs several measures, such as reducing risks and vulnerabilities, and appropriate planning and adaptation measures for the affected households. Long-term management for displaced people needs to be incorporated into an all-encompassing strategy for addressing the human security concerns. Individuals impacted by river bank erosion may encounter a multitude of human rights challenges. As an illustration, the

**Image 4.1** One women as respondent observing the future of her residence standing at river bank [Consent granted]. *Location* Hatpachil, Koizuri Union, Sirajganj District [Photo credit: The author]

safety and security become jeopardized; for instance, gender-based violence (GBV), unequal access to government support in case of basic goods and services, exploitation of the specific group of population. These group of population are subjected to abuse, neglect, and exploitation; family units are disconnected, which is especially detrimental to children, the elderly, and individuals with disabilities who may depend on familial support for survival; and they suffer the consequences of loss or damage to their family and friends. Because of riverbank erosion in Bangladesh, the affected areas that are already enduring great scarcity, including poverty, are at risk.

There is absence of reliable data and information on displaced people in Bangladesh, therefore, it appears difficult to reach a definitive conclusion. However, human security aspect is important for the formulation of future policy design and planning with respect to risks and vulnerability due to resultant river bank erosion in Bangladesh (Walsham 2010). Because of riverbank erosion, people's livelihood were disrupted by a lack of food, shelter, a decline in the sources of income, difficulty accessing public services, and disruptions in communication and transportation services. The food security situation fluctuated, for example, from January to June, the participants were able to obtain nutrition, but from July to December, their situation gets deteriorated (GOB 2000; WB 2015; GOB 2011). However, there are various coping strategies identified like engaging part-time work, performing house renovations, and temporarily living on embankments or highland (Alam 2016; Rahman and Islam 2017).

At present, Bangladesh holds the seventeenth position with the highest number of Internally Displaced Persons (IDPs) in the world (Rahman and Islam 2017). IDPs residing in Bangladesh have encountered various circumstances and encounters as a result of their socioeconomic and geographical causes, as well as the diverse events like riverbank erosion. While some are temporarily displaced because of natural disasters, IDPs who are displaced as a result of the same cause are dispersed and may not be confined to a single geographic area as in other countries. It happens for displaced people to move away from erosion areas. It was found that nearly 67% of the respondents in the study areas has moved at least once or twice. In comparison to less eroded and impacted places, this research area had a disproportionately high number of moved residents. Based on this field study, this study found that the reason behind migration are presented in Table 4.2.

People choose to relocate from one place to another, although it is not easy process at all. Vulnerable people try to get support from close relatives and friends, they try to acquire minimum land for farming, and a variety of work options, and opportunities for education, and access to basic necessities, and then affected households turn to the leave their places. Image 4.2 depicts people migrating by boat from areas affected by riverbank erosion to safer locations.

IDPs experience several unavoidable challenges at various stages of relocation. This is the most catastrophic of all natural disasters. They are vulnerable due to displacement in case of livelihood patterns and psychological distress. It is found that riverbank erosion victims became landless migrants, and that IDPs is a significant contributor to nation's recurrent poverty. Further, vulnerable people are compelled to relocate or resettle in vulnerable areas, if there is no opportunities and support available for them. The river bank erosion, the underlying reasons of the

**Table 4.2** Reasons of relocation of the vulnerable people

| Reason of relocation of the IDPs | % of the respondents |
|---|---|
| As their house floated | 21 |
| A lack of agricultural land or their original occupation | 35 |
| A lack of work opportunities | 26 |
| A desire for a safe or better quality of life | 16 |
| Other factors (fear, qualifying for job, study, marriage etc.) | 2 |

**Image 4.2** Relocation of the affected individuals by boat Location: Hatpachil, Koizuri union, Sirajganj District [Photo credit: The author] Consent wasn't taken, resulting in individuals in the boat being excluded from the image

economic losses, occurs gradually but has long-lasting effects. According to a study, two-thirds of people in the river bank erosion areas relocate permanently due to the support they receive from their neighbors, and relatives (Rana and Nessa 2017).

## 4.5 Livelihood Changes

The research examines the diverse livelihood alterations encountered by communities impacted by this unyielding natural phenomenon. Riverbank erosion is an ongoing and ancient phenomenon, influenced by the alteration and diversion of the three major rivers: the Ganges, the Brahmaputra, and the Meghna, along with their tributaries (Hossain 1993). The sediment discharge from the Bengal basin, approximated at 1050 million tons per year, ranks among the highest globally, hence intensifying the problem (Bhuiyan et al. 2017). The effects of riverbank erosion on the lives of impacted communities are multifaceted. Crop losses and property damages are among the most immediate and visible repercussions, leading to considerable economic distress for households.

Floods, sometimes associated with riverbank erosion, can significantly affect agricultural production, resulting in variations in labor demand and subsequent alterations in agricultural wages (Banerjee 2007). Households with few resources located in riverine areas are especially susceptible to these effects, as their closeness to the river increases their exposure and diminishes their resistance (Alam et al.

2019). Consequently, many rural households are compelled to make challenging choices, frequently choosing to relocate either temporarily or permanently.

River bank erosion has a negative impacts on the survivors' social, economic, and psychological existence, which can influence their behavior (Bhuiyan et al. 2017; Karim 2014). A significant number of respondents in the study area is displaced annually and they are deprived of their home, crops, and material possessions, in addition to food insecurity, economic crisis, and psychological distress, they relocate to a new location to establish a new community. This makes them environmentally and socially vulnerable in every way. The majority of respondents are affected, either it directly or indirectly. Mainly, agricultural productivity is gradually declining due to the loss of agriculture land due to river bank erosion and its effects. For instance, individuals lost their residence, leaving them with limited income opportunities. Consequently, they are constantly threatened by shelter crisis and security for their family members. To cope with these stressful situations, the affected individuals employ coping mechanisms such as relying on high rate loan or take less nutritious food and as a result their livelihoods quality also decreasing.

Local government authorities have instituted some assistance programs, such as food aid and social safety nets, but the government's institutional resources and capacities are insufficient and inadequate. However, a little number of people get that kind of support and only those who are linked with the influential society members of the community. Further, there are no industrial or development projects in this area because of the erosion risk. However, there were also fewer people employed in the agricultural industry and this industry is the one most exposed to erosion, which accounts for this circumstance. River bank erosion has caused numerous socio-economic issues and compelled residents to relocate elsewhere in search of a living (Das 2011). In an another study researchers found that socio-economic effects on livelihood are classified as many ways like loss of shelter, crop land, employment, and assets, direct effects on the people's livelihood, and indirect effects on public health and development projects (Momen and Ferdous 2023; Das et al. 2014).

It is found that nearly one-tenth of riverbank erosion prompted vulnerable people to relocate to near town or city in search of a sustainable livelihood (Hossain 1984). At the same topic but in a different research approach, it is also found that the IDPs is a social process influenced by economic developments and can modify technology and production organization, reducing rural labor needs (Saint and Willam 1980). It is also found that the affected people suffers due to absence of institutional response in formulating an effective rehabilitation and reconstruction design and its strategies (Mili et al. 2013). However, from the field study, 60% of households are experiencing economic hardship, and 85% of them have changed occupations as a result of this catastrophe. Housing, education, and medical care are among the fundamental necessities that are severely lacking in these regions. 55% of the population did not receive relief. Moreover, 98% of respondents perceive that such catastrophes can be prevented by building embankments.

This research found that the majority of the respondents were forced to shift from affected areas to urban areas and changed their livelihoods with huge difficulties. But few of them also replied that this was an opportunity for them to live with better livelihood as they get more facilities specially education and health system. River bank erosion mostly affects the poor and small landowner's household's risky lives, thus affecting their livelihood in the context of safety, shelter, and their sources of income (Brouwer et al. 2007).

### *4.5.1 Specific Vulnerable Groups Affected*

Children are vulnerable to exploitation, and abuse during the crisis period of shifting family households. Due to the fact that parents may be preoccupied with collecting aid and maintaining livelihoods, children are frequently left unattended for extended periods of time, which may increase their susceptibility and exposure to dangers. Further, seniors' citizens, young adolescent girls and individuals with disabilities encounter difficulties when attempting to abandon their shelters. The respondents said that most of the cases the vulnerable groups of people and their vulnerabilities ignores for the pressure of livelihood relocation. First of all, they fall the problems for food, safe drinking water, sanitation, health conditions of displaced households declined throughout the duration of their moving period. Despite the enhanced availability of local healthcare facilities, the standard of service provided remains unsatisfactory. Particularly in river side regions, there is a severe crisis regarding the accessibility of safe drinking water. The majority of the respondents use unsafe water for their daily needs. It is found that 12% of the total respondents feel the importance of basic health support and crisis of safe drinking water.

Riverside erosion caused a great deal of stress for riverbank residents, who lost their homes, agricultural land and production (Haque 1988). The increasing effects are income erosion, which causes households to relocate away from the area of origin, and low expenditures in the food consumption, education, and health services. However, the economic circumstances of the affected households of Bangladesh force them to maintain a precarious standard of living, and riverbank erosion exacerbates their precariousness (Uddin and Basak 2012).

## 4.6 Land Grabbing and Requisition

Bangladesh, a densely populated country in South Asia, has been coping with the intricate issues of land grabbing and riverbank erosion, significantly impacting the livelihood, food security, and overall welfare of its population. Bangladesh's physical and hydrological characteristics, characterized by low-lying coastal areas and extensive river beds, render the country especially susceptible to riverbank erosion and various natural disasters. In addition to natural pressures, human activities such as excessive groundwater extraction, canal and dam construction, and climate change effects, including rising sea levels, have all intensified riverbank erosion and its related negative consequences (Hossain 1993; Chakraborty et al. 2019; Rabbani et al. 2019).

The land ownership pattern by gender in Bangladesh does not have appropriate data. Mostly, the male households are the owner of the land properties either it house land or agricultural land. Landless households are identified as those in which no land is owned and they do not own any homestead land or cultivable land. Notably, as a patriarchal society, the majority of land has been owned by fathers or males, very few agriculture land and households land are named on female. In this research only 3 percent female was found as land owner and they receive it from their parental resource. In case of female named land and other properties become more threat not to reshape as previous status. Few respondent highlighted that after the riverbank erosion the female owner are not very active to recover their land when they had the opportunity to get it from char land. However, it is very ease to grab the char land when the female headed people are involved there.

It is important to support the households, who are struggling with poverty, relocating to recently formed char land. However, char land are exceptionally susceptible to riverbank erosion, and people are displaced due to various factors, for instance, a decline in the economy, an increase in land dispossession, and rural insecurity (Das 2011). In another research, Alam (2016) focuses on the char land, which has the potential to resolve the issue of landlessness through a planned land distribution program. New land that may emerge due to riverbank erosion in other areas require appropriate surveying to be brought under government control. It is demonstrated how a handful of land grabbers control over char land through the use of targeted violence. This research also found some instances in which a government-supported program for the allocation of khas land to the landless families were threatened by immoral Jotedars, local politicians, and dishonest officials.

## 4.7 Loss of Infrastructure

Due to lack of adequate data, the local office does not contain the exact numbers of infrastructure damage due to river bank erosion, although a preliminary estimate was derived from local office. The local economy and strategic effect of the study area's residents are severely hampered by the loss of infrastructure. Local offices, roads, embankments, markets, educational facilities, mosques, temples, and cemeteries are among the significant infrastructure losses.

### *4.7.1 Position of the Infrastructure*

Sustainability to infrastructure is dictated by the structure of the home. Along the rivers prone to erosion, the author observed a significant number of the homes consist of Kutcha, which are constructed with vulnerable materials such as corrugated tin-shade roofing, and tin walls. Mud-constructed houses (Kutcha) are especially susceptible to flooding and excessive precipitation. Mud floors, which comprise the majority of dwellings, are vulnerable to erosion when the composition of the soil structure is compromised due to inundation. People of this areas are much aware that they have uncertain to live their land areas. Very few houses are found as Pacca (brick) house and it was constructed earlier when the river was far away of that particular position. Further, the people who prepared their house for the second time after suffering mostly do the Kacha house for survival.

### 4.7.2 Essential Infrastructure

Riverbank erosion poses an obstacle to the progress of essential infrastructure. Schools and healthcare centers, and other important infrastructure, is located in high-risk zones and be vulnerable to loss or damage if no mitigation measures are implemented. Likewise, damage to power lines disrupt and disconnect daily life in the affected areas. Due to riverbank erosion, roughly 20 km of power line in Shahajadpur Upazila, Sirajgonj District, have been damaged in recent years. Recently, the Palli Bidyut Samity (Rural Electrification Board) eliminated its poles from the affected areas Because of the potential for harm. The area impacted by the disruption was isolated from electrical infrastructure. In addition to the nearby hospitals and adjacent areas, the elimination of electricity poles resulted in the disconnection of electricity connections. To continue the economic growth in the study area, the government established new roads, bridge and culvert. For education treatment and development of social activities, there are huge schools, hospital, community clinics, fire service and masjid and madrasah.

### 4.7.3 Roads and Embankment

Due to extensive riverbank erosion, Chauhali Upazila is of the most severely impacted area in Sirajganj district. Bangladesh Water Development Board (BWDB) has some projects to protect the Shahajadpur upazila from erosion of Jamuna river, the government established protection embankment in five unions, including Koizuri, Jamaitta, Jamalpur by financial investment at first phase 127 crore taka and in second phase 150 crore taka (2017–18 financial year). (BWDB field data). However, the most impacted areas of every erosion attack are the roads, bridges, and culverts among mud roads (Table 4.3).

The government has many proposed program. During the data collection period, among them the following proposed program is identified. One of the proposed project that the researcher found, relating to river bank erosion measures (Table 4.4).

**Table 4.3** Field study [data collected from BWDB official]

| |
|---|
| **Implemented Project** |
| Name of the Project: Jamuna Meghna River Erosion Mitigation Project (JMREMP) |
| Location of Project: Shahajadpur, Sirajgonj |
| Project (Start–finish): 2010–2011 |
| Main scheme of the Project: length of flood protection embankment: 10 km (Sirajgonj Part) |
| Right side of Jamuna River Embankment: 10 km (Sirajgonj Part) |
| Objective of the Project: To establish the protection project from Hatpachil to Benutia (Shahajadpur, Sirajgonj) from Jamuna riverbank erosion. |
| 1. To protect the project location from prior and seasonal flood devastation as the flood frequency increasing for climate change. |
| 2. To enhance the socioeconomic development by the facilitation of communication development. |
| 3. To secure the increase of agriculture and fish cultivation. |
| 4. To sustain the legacy of economic development by protecting the traditional weaving industry. |
| 5. To protect and develop the life and livelihoods of the local people. |

**Table 4.4** Field study [data collected from BWDB official]

| |
|---|
| **Project**: The right side of Jamuna river protection (From Brahmon Gram to Hatpachil) under Shahajadpur Upazila, Sirajganj |
| Location of the project: Shahajadpur, Belkuchi and Chauhali Upazila at Sirajganj |
| Expenditure of the Project: 64749.12 lac taka |
| Project start: July 2020 |
| Project closed: June 2023 |
| Scheme of the Project: River embankment protection [Rehabilitation of Spar; Embankment Repairing; Protection work rehabilitation and others] |

## 4.8 Income and Occupational Patterns

The monthly income of displaced people is not easy to properly calculation because of the nature of employment, such as irregular and seasonal migration, and the absence of information regarding outstanding loans. Displaced people does not maintain financial records of their earnings. The field study showed that the number of earners per displaced household was 1.85, which was greater than the national average of 1.27 for rural households in Bangladesh (BBS 2010). A primary occupation of 58 percent of the respondents in this study was agriculture and fishing which served as their main source of income. Undoubtedly, riverbank erosion has had an effect on employment. An increasing number of fishermen are experiencing marginalization due to displacement. Because of relocation in different areas, households are more susceptible to various internal and external challenges that are inherent in migratory labor (Table 4.5).

Table 4.5 Monthly income distribution of adjunct household

| Income (BDT) | No. of households |
|---|---|
| Less than 4000 | 7 |
| 4001–6000 | 10 |
| 6001–8000 | 14 |
| 8001–10,000 | 19 |
| 10,001–12,000 | 9 |
| Over 12,000 | 7 |

### 4.8.1 Women Occupation

Mostly women are not engaged with commercial job but few of them are involved with government job like teaching in primary school and serving in health care institutions. Among 27 female respondents, this study found only 3 of them are involved with service sector and few are participating in household activities and other seasonal work. Particularly women are engaged themselves for daily household activities and handicrafts works. But from their view, women can earn money and they can buy assets and land to avoid their future vulnerabilities, although few of them earn some money after the households' head permission.

## 4.9 Educational Institutions

Generally, disasters prevent children from attending schools. Consistent absenteeism increases the likelihood of eventual drop out in schooling. Depending on the geographical context, riverbank erosion may cause structural harm to educational institutions, resulting in the subsequent loss of educational institutions and materials. Due to riverbank erosion, the educational institutions of the impacted villages were primarily moved.

**Case Story 1**

Md. Yakub Ali, Pakurtola Primary high School, Jalalpur, Upazila; Sahajadpur

Pakurtola primary school drawn in the 2022 riverbank erosion. Before the erosion, the total students' number was 150, but after the erosion decrease the number 60. 90 students were dropped out. After the erosion, teachers have managed their teaching at different houses. Now it is a tin shade building and the number of the students' attendees 25-30 per day. Rest of the students are now going to fishing with their father. Their parents are not much aware about this issue and the rate of drop out is still increasing because of riverbank erosion.

**Image 4.3** River embankments. *Location* Jamuna River, Saidabad Union, Sirajganj District [Photo credit: The author]

A riverbank erosion resulted in the destruction of four primary schools and one secondary school throughout the study period. 60 girls and boys between the ages of nineteen who were enrolled in school dropped their studies following displacement. A total of twenty-one students were involved in child labor out of 66 households as a means to support their families. The situation intensifies the practice of child marriage. Girls are entering into matrimony at the early age of fifteen. The majority of high schools are situated at greater distances from the residences of displaced individuals as they are enrolled the previous one. The girls who are predominantly adolescent are completely avoid the school though few of them goes for months. The majority of respondents who have lost their ancestral homes have relocated, so their children cannot attend the same school.

Image 4.3 illustrates the embankment that protects against riverbank erosion in the study location.

Furthermore, the availability of conveyance vehicles is limited. In general, social insecurity, ignorance regarding the importance of formal education, and religious convictions are the primary contributors to the rising high school dropout rate. Approximately 75% of respondents state that river erosion has tremendous effects on education of their children. 50% of the respondents are heavily affected, and 91% of them have relocated (Image 4.4).

**Image 4.4** A vulnerable man sat on the riverbank. The embankment of his land area had already collapsed into the Jamuna River. [Consent granted]. *Location* Hatpachil, Koizuri Union, Sirajganj District [Photo credit: The author]

## 4.10 Religious Institutions

Among the respondents, 94% of the population is Muslim, and 6% is Hindu. The religious institutions are devastated due to erosion and the mosque and temples was washed away. There have been around 21 mosques and 7 temples found to have deteriorated. Although some of them were reconstructed, their aesthetic beauty was not recovered during the study period. Many people experience the loss of their parents' and grand parents' graves. Few madrasah was also devastated for the reasoning of riverbank erosion. This creates huge vulnerabilities because mostly the orphanage boys are the students of this religious institutions. After devastating this kind of institutions the students suffer most and the eroding site poses an emotional risk to that affected people.

## 4.11 Market and River Bank Commodities

In Shahajadpur Upazila, which is located on the bank of the Jamuna River, there are numerous well-known bazaars. At the Upazila sadar (center), there were hundreds of busy sellers; thousands of small vendors depend on the Upazila market for their livelihood. At least 150 affected business people were leaving thousands of vendor capitals. The largest shop, according to the locals, lost during that time in less than 30 minutes, killing 70 people there, in addition to losing 200 stores. However, bank erosion strength differs significantly between rivers due to a number of factors, for instance, river bed material, fluctuations in the level of water, flow velocities near the bank, the river's platform, and the water and sediment supplies that enter the river. For instance, recently dropped, loosely compacted bank materials comprised of sand and sediment are extremely susceptible to erosion.

## 4.12 Environmental Losses

The earlier studies demonstrate the connection between environmental degradation and migration, including riverbank soil erosion and migration (Das et al. 2013), and riverbank erosion and migration (Kamal and Abedin 2019). However, infrastructure and land is destroyed by erosion, and this have also short and long-term consequences on environment (Walsham 2010). The flora, fauna, and fisheries sectors are all negatively impacted by the river erosion (Uddin and Basak 2012). In the research area, tree plantations are a fairly popular and regular practice surrounding home, and every home has large indigenous trees. However, the trees are no longer there due to extensive erosion. When affected people realized the land might be deteriorated, beforehand many people sold their young trees. Further, as a consequence and huge demand of people's settlement of the rising population, deforestation is occurring in order to provide for the expanding requirements of agricultural and homestead land. It is to be noted that deforestation at a faster pace results in the loosening of soil particles, which facilitates the progressive erosion of the riverbank soil. Furthermore, the area of cultivable land is declining due to erosion, and fishing industry is the one that is most influenced by river bank erosion. The siltation process has an effect on fish populations as well as river traffic, which is fast declining.

### *4.12.1 Hydraulic Forces of Its Rivers*

Particularly during periods of high flow of water, the coastal lands of Bangladesh provide little resistance to the hydraulic forces of its rivers. Large braided rivers, such as the Brahmaputra and the Jamuna, have the inherent characteristic of both eroding banks and depositing sediment on the riverbed. This process results in the formation of char islands, which are new riverine formations. These inhabited char islands, which follow the river's course, are extraordinarily susceptible to fluctuations in river conditions.

**Image 4.5** Local residents placed some sandbags to protect against riverbank erosion. [Consent granted]. *Location* Gharjan Union, Chauhali, Sirajganj District [Photo credit: The author]

### *4.12.2 Rainfall and Heavy Flooding*

A large area of the landmass is made using sediment transported by rivers in the country. Bangladesh is situated in a monsoon climate zone. The absence of adequately constructed river embankments can result in soil erosion caused by excessive precipitation and subsequent inundation. In addition, the fluctuating pattern of precipitation caused by climate change contributes to atypical inundation. In association with the increased influx of river water from the upper catchment countries, this inundation exacerbates riverbank erosion (Momen and Rahman 2023). Riverbank erosion may occur both during and after to the moving of water (Image 4.5).

## 4.13 Nutrition and Food Security

Food insecurity is one the most significant consequence in any disaster (Paul 1998). At this time, riverine families adopts a variety of coping strategies to cope with the threats. These adaptation strategies could be implemented at the grassroots level or through government policy. However, all households are not affected equally by disaster and threats, and thus the responses to the incident vary depending on their economic capacity and political and social connections (Emel and Peet 1989). It is predicted that some natural hazards including river bank erosion may have devastating effects on agriculture and food security throughout the globe, especially in developing nations such as Bangladesh (Begum and Momen 2020; Keya and Harun 2007; Islam and rashid 2011; WB 2013, IPCC 2014).

Given the prevalent malnutrition rates throughout the country, the nutrition situation could rapidly and easily deteriorate in the aftermath of a catastrophe, with children under the age of five becoming especially susceptible. The destruction of food crops caused by farmland erosion may reduce access to a variety of nutrient foods, which in turn may result in inadequate intake of energy and micronutrients. While erosion may not have an immediate and direct effect on nutrition, it can exacerbate vulnerabilities and contribute to a rise in under nutrition.

## 4.14 Losses of Aesthetic Beauty

The losses in artistic beauty cannot be clearly expressed in money. According to the interviews with locals, the area was once a never-ending paddy crops but now becomes a sandy wasteland. The vanished settlement has been moved, and it reached to the middle of the river. Further, aesthetic beauty of indigenous plant species that had been documented by the neighborhood forest office is no longer present in the study location. People change their occupations and move to new locations as the market disappears. In the study area, there are several unplanned roads and embankments that illustrate the deplorable conditions of the local area.

## 4.15 Seen and Unseen Vulnerabilities

Riverbank erosion causes substantial damage to the impacted population (Mollah and Ferdaush 2015). Each year, an enormous amount of agriculturally productive land is lost to erosion. The agricultural land's cropping pattern has been drastically altered, and it sometimes gets unfertile. In the study areas, infrastructure damages are also substantial and the affected households frequently relocate their homes. This increases the number of landless people, thus creating the surplus agricultural labor force, which generates severe unemployment in agricultural sector. All of these are the common issues and scenario of the study areas, however, the adaptation strategy and resilience mechanisms should be dynamic to deal with the crisis.

On the other hand, due to the fact of displacement the vulnerabilities are increasing continuously. Most of the respondents said that they predict some common patterns of vulnerabilities related with livelihoods, but some unfamiliar vulnerabilities are awaiting for them. One woman respondent (now living in Dhaka) replied that she was introvert lady who never went to other houses, she was also forced to work as house cleaner or garments worker in Dhaka city, although she is earning some

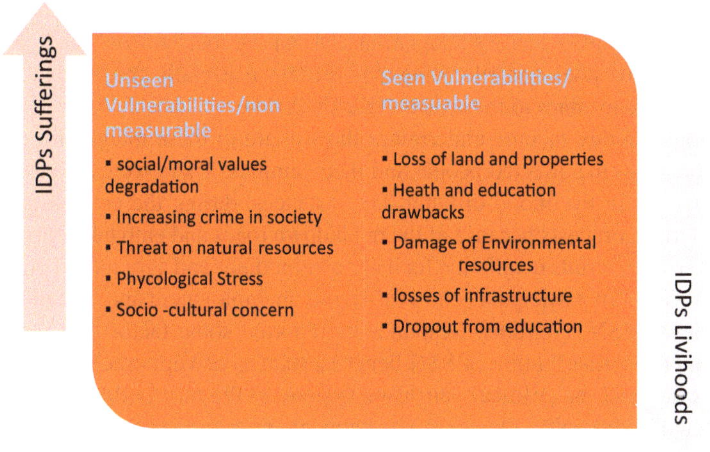


**Fig. 4.2** Seen and unseen vulnerabilities (the author)

money. In addition, children are mostly discouraged to go school regularly as their mother is not at home every day, most surprisingly the children are also attracted to get some money some roadside labor work, such as flower selling. Further, sometimes the adolescent boys and girls both have the tendency to engage themselves with the crime because of lack of their knowledge and easy access of that unauthorized profession. As a result, the social and moral values are decreasing though their household income are increasing (Fig. 4.2).

## 4.16 Voices from the Marginal Groups

With women on chars in the affected areas has shown interest in relocating to the original land and citing access to education for their children and increased safety as significant reasons for their desire. Further, due to long-distance relocation for labor, women are frequently vulnerable in various ways. It is reported from the respondents that most of the vulnerable women with children had left to Dhaka to be working in the ready-made garment (RMG) industry since the early 1990s. Moreover, many of the cases, relocation decisions are largely 'voluntary', in some instances, river bank erosion gets deteriorated and the adaptation capacity becomes weak that entire families or even entire communities may choose to relocate. It is observed that the majority were compelled to reside in Sirajganj district, such as railway colonies, slum areas, outline of flood protection embankments, and only a little number of them had received assistance from the government and others.

The erosion has negative effect on the socioeconomic well-being of riverbanks residents. As stated, affected households due to river bank erosion lose their agricultural land and assets, which impacts on their monthly income and traps them in further poverty. Although erosion relocates many households, victims in less affected areas employ various adaptation and coping mechanisms, such as selling their land and livestock and relocating to new areas, especially char land in Sirajganj district in order to adapt to the erosion (Karim 2014).

The low monthly income undermines the welfare of these marginalized groups, posing threats to the affected people and their families. There were also some other factors that make livelihood difficult. For instances, during the erosion process, it increases transport costs because of the breakdown roads and embankments and, further, this erosion results in inflation for daily commodities and agricultural production.

The majority of participants cite year-round unemployment and lack of access to government's SSN programs. Respondents reported some factors are responsible for the improper distribution of SSN benefits, such as corrupt practices of the issuing authority, and the political patronage system, to the most vulnerable groups in the affected community. On the other hand, there are other external factors, for instance, the government's inadequate relocation programs, push the displaced households at risk for further relocation.

The findings shows that a high level of inequality in receiving government supported SSN benefits to affected households, adding to the theoretical debated and discussion that displaced people through the framework of socio-environmental inequalities rather than environmental 'push' theories (Hunter et al. 2015). This section can be concluded by referring the displaced people, demonstrating how certain factors, for instances, absence of employment opportunity at local level, lack of credit support, breakdown of transportation system, and no access to improved healthcare, and sanitation facilities, have a greater negative impact on the riverine population in Bangladesh.

## 4.17 Avoidance Toward Riverbank Erosion [Respondents' Views]

***Plantation Programmes***. River bank erosion can be protected by plant, particularly roots. Strong roots strengthen the soil along riverbanks, thereby averting catastrophic erosion. Along dynamic streams, plant and vegetation absorbs the impact of heavy rainfall and flooding, thereby reducing erosion.

### *4.17.1 Tree Arrangement*

A few fallen trees provide protection along this riverbank. To prevent bank erosion, horizontally secured trees are utilized. Trees decelerate water flow that reduces erosion and entraps sediment. When tree branches capture sediment, they establish a healthy soil foundation that reduces erosion.

### 4.17.2 *Vegetated Concrete Blocks*

Stream and riverbanks are stabilized by concrete block adorned with tree plantation or indigenous vegetation. Substantiating riparian vegetation may work as shield, thereby maintaining the flow of water.

**Case Story 2**
Jahid Hasan [Student], Hatpachil, Koizuri Union

Government has taken few projects and most of the cases, it is estimated to finish at least 2/3 years, hence the respondent don't find any speedy work. He saw 5/6 houses washed away per hour, especially during the flash flood. He suggested the places of where the stream is high, the authority may introduce the "T" shape embankment, and it can reduce the erosion.

## 4.18 Conclusion

Riverbank erosion causes employment crises and land loss, so developing substantial poverty, which is worsening annually in riparian areas. There are substantial issues that include resource constraints, inadequate management and unplanned rehabilitation, as the sustaining aspects of riverbank erosion in Bangladesh. Nonetheless, it is evident that the way of life of the impacted individuals is significantly threatened by exposure to hazards and vulnerabilities, particularly the escalating erosion of riverbanks. Despite the persistent risk of losing their land and infrastructure, the majority of households in areas susceptible to riverbank erosion do not perceive this gradual threat as immediate or direct. The experiences of riverbank erosion proved catastrophic, especially for landowners who faced persistent impoverishment and livelihood concerns.

## References

Adamo SB (2008) Addressing environmentally induced population displacements: a delicate task. Background Paper. In: Population-environment research network cyber seminar on environmentally induced population displacements held on August 18–29

Alam GMM, Alam K, Mushtaq S, Sarker MNI, Hossain M (2019) Hazards, food insecurity and human displacement in rural riverine Bangladesh: implications for policy. Int J Disaster Risk Reduct (Elsevier BV) 43:101364. https://doi.org/10.1016/j.ijdrr.2019.101364

Alam GMM (2016) An assessment of the livelihood vulnerability of the riverbank erosion hazard and its impact on food security for rural households in Bangladesh. PhD thesis, School of Commerce, University of Southern Queensland, Australia, 2016

Baqee MA (1997) Coping with floods and erosion in Bangladesh char-lands. Asia Pacific J Dev 4(2):38–52

Banerjee L (2007) Flood disasters and agricultural wages in Bangladesh. Wiley 38(4):641–664. https://doi.org/10.1111/j.1467-7660.2007.00427.x

BCCSAP (2009) Bangladesh climate change strategy and action plan. Ministry of Environment and Forest. Dhaka, September 2009

Begum MM, Momen MN (2020) Drought. In: Romaniuk S, Thapa M, Marton P (eds) The Palgrave encyclopedia of global security studies. Palgrave Macmillan, Cham. https://doi.org/10.1007/978-3-319-74336-3_323-1

Bhuiyan MAH, Islam SMD, Azam G (2017) Exploring impacts and livelihood vulnerability of riverbank erosion hazard among rural household along the river Padma of Bangladesh. Environ Syst Res 6:1–15. https://doi.org/10.1186/s40068-017-0102-9

CEGIS (2010) Prediction of riverbank erosion along the Jamuna, Ganges and Padma rivers in 2010. North South University Library OPAC. https://opac.northsouth.edu/cgi-bin/koha/opac-detail.pl?biblionumber=23672&shelfbrowse_itemnumber=37605#shelfbrowser

Chatterjee S, Mistri B (2013) Impact of river bank erosion on human life: a case study in Shantipur Block, Nadia District, West Bengal. Int J Hum Soc Sci Invent 2(8):108–111. https://ijhssi.org/papers/v2(8)/Version-3/Q028301080111.pdf

Chakraborty R, Khan KM, Dibaba DT, Khan MA, Ahmed A, Islam MZ (2019) Health implications of drinking water salinity in coastal areas of Bangladesh. Int J Environ Res Public Health 16(19):3746. https://doi.org/10.3390/ijerph16193746

Das B (2011) Stakeholders' perception in identification of river bank erosion hazard: a case study. Nat Hazards 58(3):905–928. https://doi.org/10.1007/s11069-010-9698-z

Das KT, Halder HK, Gupta ID (2014) River bank erosion induced human displacement and its consequences. Living Rev Landsc Res 8(2014):3. https://doi.org/10.12942/lrlr-2014-3

Dekaraja D, Mahanta R (2021) Riverbank erosion and migration inter-linkage: with special focus on Assam, India. Environ Syst Res 10:6. https://doi.org/10.1186/s40068-020-00214-0

Elahi KM, Ahmed QS, Mafizuddin M (ed) (1991) Riverbank erosion, flood and population displacement in Bangladesh. Riverbank erosion impact study. Jahangirnagar University, Savar, Dhaka

Emel J, Peet R (1989) Resource management and natural hazards. In: Peet R, Thrift N (eds) New models in geography, pp 49–76

Ferris E (2011) Climate change and internal displacement: a contribution to the discussion. UNHCR Bellagio Roundtable, 22–26 February 2011

GOB (2011) The sixth five-year plan, 2011–2015. Government of the People's Republic of Bangladesh, Dhaka, Ministry of Planning, 2011

GOB (2000) Report of the task force on comprehensive food security policy for Bangladesh. Government of the Peoples' Republic of Bangladesh, Dhaka, 2000

Haque CE (1988) Impacts of river bank erosion hazard in the Brahmaputra-Jamuna Floodplain: a study of the population replacement and response strategy. University of Manitoba. https://mspace.lib.umanitoba.ca/home

Hossain M (1993) Economic effects of riverbank erosion: some evidence from Bangladesh. Disasters (Wiley) 17(25–32). https://doi.org/10.1111/j.1467-7717.1993.tb00485.x

Hunter LM, Luna JK, Norton RM (2015) Environmental dimensions of migration. Annu Rev Sociol 41(377–397). https://doi.org/10.1146/annurevsoc-073014-112223

Islam MF, Rashid AB (2011) Riverbank erosion displaces in Bangladesh: need for institutional response and policy intervention. Bangladesh J Bioeth 2(2):4–19. https://www.banglajol.info/index.php/BIOETHICS/article/view/9540

Islam MS, Hasan T, Chowdhury MSIR, Rahaman MH, Tusher TR (2013) Coping techniques of local people to flood and river erosion in char areas of Bangladesh. J Environ Sci Natl Resour 5(251–261). https://doi.org/10.3329/jesnr.v5i2.14827

Islam GMM, Alam K, Mushtaq S (2018) Drivers of food security of vulnerable rural households in Bangladesh: implications for policy and development. South Asia Econo J 19(43–63). https://doi.org/10.1177/139156141876107

Islam MR (2018) Climate change, natural disasters and socioeconomic livelihood vulnerabilities: migration decision among the char land people in Bangladesh. Soc Indic Res 136(575–593). https://doi.org/10.1007/s11205-017-1563-y

IPCC (2014) Climate change 2014: impacts, adaptation and vulnerability, Part A: global and sectoral aspects, contribution of working Group II to the fifth assessment report of the intergovernmental panel on climate change. Cambridge University Press, Cambridge, 2014. https://www.ipcc.ch/site/assets/uploads/2018/02/WGIIAR5-FrontMatterA_FINAL.pdf

Karim AHMZ (2014) Flood and riverbank erosion displaces: their indigenous survival strategies in two coastal villages in Bangladesh. Can Center Sci Educ 10(4):16–26. https://doi.org/10.5539/ass.v10n4p16

Keyqn MK, Harun SMR (2007) Riverbank erosion induced stress and coping of displaced women in Bangladesh. Empowerment 14:17–30

Laczko F, Aghazarm C (2009) Migration, the environment and climate change: assessing the evidence. International Organization for Migration, Geneva. https://reliefweb.int/report/world/migration-environment-and-climate-change-assessing-evidence?gad_source=1&gclid=Cj0KCQjwh7K1BhCZARIsAKOrVqEVCpQsVRfGKRqdeZnoXKljfJ5S4E1ww4PpH1yYg0q2jhmMftx3jRAaAnzVEALw_wcB

Lein H (2009) The poorest and the most vulnerable? On hazards, livelihoods and labelling of riverine communities in Bangladesh. https://doi.org/10.1111/j.1467-9493.2008.00357.x

Mahmood R (1986) Bangladeshi returned migrants from the middle east: process, achievement, 157 pp. https://books.google.com.bd/books/about/Bangladeshi_Returned_Migrants_from_Middl.html?id=2Du0zwEACAAJ&redir_esc=y

Mili N, Acharjee S, Konwar M (2013) Impact of flood and river bank erosion on socio-economy: a case study of Golaghat revenue circle of Golaghat District, Assam. Int J Geol Earth Environ Sci 3(3):180–185

Mollah TH, Ferdaush T (2015) Riverbank erosion, population migration and rural vulnerability in Bangladesh (a case study on Kazipur Upazila at Sirajgonj District). Ecol Environ Res 3:125–131. https://doi.org/10.13189/eer.2015.030502

Momen MN, Ferdous J (2023) Human resource capacity and changes in public sector organizations in Bangladesh. In: Governance in Bangladesh. New frontiers in regional science: Asian perspectives, vol 68. Springer, Singapore. https://doi.org/10.1007/978-981-99-0424-2_4

Momen MN, Rahman MM (2023) The impoverishment of political discourse and legal rights in transboundary Teesta water management. World Water Policy 9(1):11–28. https://doi.org/10.1002/wwp2.12093

Paul BK (1998) Coping mechanisms practised by drought victims (1994/5) in North Bengal, Bangladesh. Appl Geogr 18(4):355–373. https://doi.org/10.1016/S0143-6228(98)00026-5

Rabbi H, Saifullah ASM, Sheikh MS, Sarker MMH, Bhowmick AC (2013) Recent study on river bank erosion and its impacts on land displaced people in Sirajgonj riverine area of Bangladesh. World J Appl Environ Chem Int J Appl Environ Sci 2(2):36–43

Rahman MM, Islam MN (2017) Bank erosion pattern analysis by delineation of course migration of the Padma River at Harirampur Upazila using satellite images and GIS part II. J Geol Geophys 2017(6):1–6. https://doi.org/10.4172/2381-8719.1000284

Rabbani G, Munira S, Saif S (2019) Coastal community adaptation to climate change-induced salinity intrusion in Bangladesh. IntechOpen. https://doi.org/10.5772/intechopen.80418

Rana MS, Nessa AM (2017) Impact of riverbank erosion on population migration and resettlement of Bangladesh. Sci J Appl Math Stat 5(2):60–69. https://doi.org/10.11648/j.sjams.20170502.11

Rana MS, Nilufa A, Kabir H, Asmaul H et al (2022) Consequences of riverbank erosion among rural household along the river of Meghna in Bangladesh. J Environ Hazard 6(2022):172. https://doi.org/10.37421/2684-4923.22.6.172

Saint WS, Willam GW (1980) Cropping systems structural changeand rural-urban migration in Brazil. World Development Press Ltd. 8(3):259–272

Salehin M et al (2018) Mechanisms and drivers of soil salinity in coastal Bangladesh. In: Nicholls R, Hutton C, Adger W, Hanson S, Rahman M, Salehin M (eds) Ecosystem services for well-being in deltas. Palgrave Macmillan, Cham. https://doi.org/10.1007/978-3-319-71093-8_18

Uddin MN, Rahman MM (2011) Socio-economic impact of erosion along the right bank of the Jamuna River in Bangladesh. DUET J 1:35–42. https://www.researchgate.net/publication/266069473_Socio_-Economic_Impact_of_Erosion_along_the_Right_Bank_of_the_Jamuna_River_in_Bangladesh

Uddin AFMA, Basak JK (2012) Effects of riverbank erosion on livelihood. Unnayan Onneshan, Centre for Research and Action on Development. https://doi.org/10.13140/RG.2.2.24933.37609/1

Walsham M (2010) Assessing the evidence: environment. Climate change and migration in Bangladesh, International Organization for Migration (IOM). https://publications.iom.int/system/files/pdf/environment_climate_change_bangladesh.pdf

WB (2013) Turn down the heat: climate extreme regional impact and case for resilience. A Report for the World Bank by the Potsdam Institute for Climate Impact Research and Climate Analytics. World Bank, Washington DC, USA. https://www.worldbank.org/en/topic/climatechange/publication/turn-down-the-heat-climate-extremes-regional-impacts-resilience

WB (2015) World development indicators, the World Bank, Washington, DC, USA, 2015. https://documents.worldbank.org/en/publication/documents-reports/documentdetail/795941468338533334/world-development-indicators-2015

# Chapter 5
# Avoidance of Riverbank Erosion in Bangladesh: Capacity and Performance

**Abstract** Bangladesh is mostly an agriculture-based country that is subject to natural calamities, making riverbank erosion worse. Historically, river bank erosion directly and indirectly affects more than one million individuals each year, most of whom become landless along the river. Despite riverbank erosion's harmful effects on livelihood, no significant efforts have been made to reduce it. People displaced by riverbank erosion do not receive a significant response to their adaptation measures. However, this chapter have drawn the findings that the displaced population receives no significant support from the GoB in securing livelihood, especially in reconstructing homes or opportunities in income generation. A meticulously selected direct interview was conducted with 66 individuals from 54 families, including 20 households from urban slums in Dhaka, FGDs, KIIS, and case studies. Detailed justification of data collection is provided in chapter one.

## 5.1 Introduction

Riverbank erosion impacts millions of lives, is in many cases, and unseen vulnerabilities, and disaster plan should include rescue, rehabilitation, and protection measures to reduce the vulnerability (Rabbi et al. 2013). However it is observed that mainstream media does not cover regular news and it is not a priority agenda for the government attention. The media may act a significant role in demonstrating the impacts of river erosion by producing documentaries and reporting on the subject. However, the vast majority of them merely exhibit the livelihood vulnerabilities of the affected households, rather focus on the news and updates of the river bank erosion, which is not acceptable at all. The findings of this study highlight that riverbank erosion caused land loss, and driven relocation. Affected households experienced numerous livelihood vulnerabilities, but received little institutional

 
M. Begum, *Understanding River Bank Erosion in Bangladesh*, Springer Water,
https://doi.org/10.1007/978-3-032-00288-4_5

assistance. Government assistance was at best minimal. Participants in the study depend on other family members and social networks to mitigate the effects. From the field study, neither the general public nor the organizations responsible for aiding participants listened to the participants' voices and concerns. In contrast, it is found that affected community is somehow engaged in activism to reduce the sufferings (Momen et al. 2020). River victims typically receive little aid, and many are forced to migrate to Dhaka's overcrowded slums. Thus, the calamity impedes national development. It is a natural disaster that frequently threatens the livelihood of those who live along the river (Mehedi et al. 2010; Shamsuddoha and Chowdhury 2007). When it comes to property losses in Bangladesh, riverbank erosion is regarded as the worst calamity, since it is slow, silent, and ranks first (Rahman et al. 2015).

Due to the law of rivers, one bank erodes while the other bank rises. However, there is an increase in char land that can benefit river-eroded individuals. Therefore, it is crucial to use and distribute the char lands among the affected households. The Chars Livelihood Programme and the International Federation of Red Cross and Red Crescent Societies' (IFRC) community-based disaster risk reduction programme, where affected families due to river erosion are a specific vulnerable community. These organizations have few programs that attempt to benefit to those livelihood along the main rivers in Bangladesh. Undoubtedly, assisting households in reestablishing their homes on improved foundations, assisting them in enhancing their means of subsistence, and enhancing their access to fundamental services is unquestionably a crucial endeavor (Momen and Ferdous 2023a), given that affected households still prefer to remain in the affected area, it is likely that many of those displaced will be benefited from such programs.

Although there are some coping and adaptation strategies of households, but the affected households generally agree that those directly affected by river erosion have received insufficient policy attention and resource allocation. This is likely due in part to the difficulty of effectively managing a geographically dispersed population. However, there have been some recent efforts of providing improved information and national projections to communities on risks, such as the CEGIS/UNDP pilot program using flags to identify erosion-prone areas, are encouraging. However, these initiatives make little progress to prevent relocation. In this context, relevant policies and programs are required to aid displaced people (Walsham 2010), particularly to those who are compelled to relocate and frequently wind up in urban slums in Dhaka city in Bangladesh.

## 5.2 Policies, Legal Frameworks and Initiatives

It is common knowledge that "internal displacement" is a complex issue, and numerous organizations are devoted to addressing its various facets. In accordance with the Standing Orders on Disaster (SOD), adopted by the MoDMR, is tasked with the evaluation of vulnerabilities and risks with the purpose of developing Risk

Reduction Action Plans (RRAP) and the strategies for their execution. The Table 5.1 specifies that the Action Plan must be in accordance with the strategic plans of the country.

## 5.3 Legal and Institutional Framework

The Acquisition and Requisition of Immovable Property Ordinance II (1982) and its amendments (1989/93/94) and other land related laws and administrative manuals related to allusion/delusion land, char, and khas land administration in Bangladesh govern land acquisition (BWDB 2014). The 1982 Ordinance emphasizes the compensation for the damages of ancestral homes and assets, such as agriculture crops, trees, and livestock, however, this legal instrument does not provide for resettlement or livelihood restoration for affected households/businesses.

There are various kinds of mitigation measures for the affected people, for instance, the ADB's integrated safeguard policy statement (SPS) intends to reduce relocation and requires time-specific strategies to increase income and to improve the quality of life for persons that can be applied to the affected people by river bank erosion. The project land purchase and rehabilitation principle has been aligned with ADB's SPS because the Ordinance, adopted in 1982, fails to meet various safeguard rules (BWDB 2014).

### *5.3.1 Income and Livelihood Restoration Program (ILRP)*

Through river erosion, impoverished households not only lose their land and assets, but also experience socioeconomic vulnerabilities. Before the situation further worsens, it is important to adopt the necessary steps to encounter erosion and guarantee the victims' livelihoods. In addition, a variety of measures should be adopted at various levels to reduce river erosion losses. In this respect, ILRP for the impoverished, especially for poor female-headed households. The ILRP provides training for developing human resource and occupational skills and financing for operating businesses in the affected areas.

### *5.3.2 Resettlement Program Implementation Institutional Arrangements*

The BWDB serves as the project's executing agency. BWDB forms a Project Management Office (PMO), headed by a Project Director (PD), to carry out the Project. The PMO establishes a Resettlement Unit (RU) which operates under the

**Table 5.1** Policy framework on disaster management in Bangladesh [adapted from various strategic plan in Bangladesh]

| | |
|---|---|
| National Adaptation Programme of Action (NAPA) 2005 | NAPA underscores the detrimental consequences of climate change, such as extreme events and variation, and its objective to advance sustainable development in Bangladesh |
| Climate Change Strategy and Action Plan (2009–2018) | The strategy and action Plan is built on six pillars:<br>(a) Food security, social safety and healthcare; (b) Comprehensive disaster management programme; (c) Infrastructure; (d) Research and development (R&D); (e) Mitigation and low carbon emission; (f) Capacity building and institutional framework |
| National Plan for Disaster Management (2010–2015) | For the purpose of minimizing the risks and vulnerability to natural and human-induced disasters to a manageable and acceptable level. Community-level disaster preparedness, disaster response, recovery, and rehabilitation were the primary objectives of the national plan. In addition, this national plan has addressed the mainstreaming of disaster risk reduction and climate change adaptation in all sectoral policies and plans, as well as the mitigation measures for any new hazard and the frequency of these threats |
| Disaster Management Action Matrix (2010–2015) | The strategic objectives, key achievable targets, expected results, agenda for implementation, lead agency/department, and supporting agency/department were all outlined in the action matrix. Unfortunately, this plan fails to incorporate protection measures for people affected and the vulnerabilities of displacement |
| National Health Policy (2011) | This policy denotes the provision of primary and emergency healthcare to all individuals. Quality healthcare necessitates nineteen specific objectives. Accessibility for underprivileged individuals, gender equity, immunization, family planning, mobile clinics, climate-change-induced diseases, and e-health have all been addressed |
| National Education Policy (2010) | Quality of education, ethical learning, and mainstreaming education for displaced children are particularly prioritized in education policy |

(continued)

**Table 5.1** (continued)

| | |
|---|---|
| Sixth Five Year Plan (2011–2015) | This plan emphasizes that underprivileged people have restricted access to land, including char lands, khas lands, and water bodies. Nevertheless, there are specific land laws and policies that are designed to distribute such land to the landless and vulnerable. However, the interests of the impoverished are rarely safeguarded during the distribution process. Despite the introduction of certain plans of action, such as the recovery and rehabilitation of the affected people as a result of climate change, the development and relocation initiatives and schemes are extremely slow and may take a longer time to reach displaced people |
| Seventh Five Year Plan (2016–2020) | The Seventh Five-Year Plan (2016–2020) of Bangladesh emphasized disaster management as a key area for achieving sustainable development. These efforts aimed to reduce losses from disasters and support sustainable development |
| Eighth Five-Year Plan (2020–2025) | The Eighth Five-Year Plan (2020–2025) of Bangladesh highlights disaster management to address climate change and disaster risks. This plan aims to enhance resilience, reduce vulnerabilities, and ensure sustainable development despite increasing disaster risks |
| National Land Use Policy, 2001 | In order to prevent land degradation and ensure its optimal utilization, the policy emphasizes land use policy. Nevertheless, there are three distinct categories of khas lands: agricultural, non-agricultural, and water bodies |
| Char Development and Settlement Project (CDSP), Gucchagram (rehabilitation project for climate affected community), Asrayan (poverty alleviation programs through rehabilitation and income generation activities), Abashan (rehabilitation project), Grihayan (housing for homeless and rehabilitation of slum dwellers & low income peoples in multistoried building on government land in Dhaka city | IDPs experience many challenges in accessing char lands for their resettlement, and distribution of khas lands to marginalized community is always in question, therefore, these initiatives are intended to support for the displaced people |
| The 1998 Guiding Principles on Internal Displacement' by United Nations High Commissioner for Refugees (UNHCR) | The guiding principles emphasize to the specific needs and vulnerabilities of IDPs worldwide. The Guiding Principles protect the civil rights and provide support for IDPs as well as voluntary and safe return, resettlement and reintegration |

supervision of the Chief Resettlement Officer (CRO). This unit is in charge of executing the Resettlement Program (RP), which involves disbursing compensation and rehabilitation benefits, with support from the Resettlement Implementing NGO (RP-INGO) and the relevant BWDB office. RUs is tasked with executing an ILRP, with the support of a national Livelihood Development Non-Governmental Organization (LD-NGO).

## 5.4 National Response and Local Adaptive Capacity

Significant areas of Bangladesh persistently remain very susceptible to the possibility of flooding in future, and the impact of resultant climate change may further intensify the problem over the 21st century. In the future, a significant number of individuals will remain at risk of flooding. Understanding patterns of migration from flood-affected areas, including those caused by extreme floods, seasonal flooding, and extended water-logging, is crucial for developing specific strategies that can minimize risks and vulnerabilities and provide benefits for affected households. Nevertheless, the study areas are prone to erosion often face recurrent displacements, which present a substantial hindrance to the successful implementation of long-term rehabilitation and recovery efforts.

Rapid action is necessary in response to any disaster with interventions usually involving immediate intervention. The local government can enhance relief distribution by collaborating with other development partners, for instance, national government, NGOs and the private sector. Typically, in response to a disaster, the government provides rice and currency to the affected people. However, effective coordination between the local government and Disaster Management Committees (DMCs) is necessary to identify and help the recipients.

There has also been an increase in community involvement in places affected by river bank erosion in recent years. The development of bank protection structures at Sirajganj has influenced the changes in the width of the river. This specific kinds of infrastructure hinders the river's ability to widen because of bank erosion. Additionally, it has been stated that there is currently no nationwide erosion forecasting system in place.

## 5.5 Institutional Responses Addressing Riverbank Erosion

Typically, institutional responses to riverbank erosion encompass three main approaches: (1) Providing immediate relief in the form of essential supplies like food, medicine, clothing, and financial assistance; (2) Implementing protective measures, including riverbank dams, dredging riverbeds, and placing layers of stones to mitigate erosion; and (3) Supporting affected individuals in finding new employment opportunities when they are forced to relocate. In the early 1990s, it is documented that in Bangladesh, the government and non-governmental organizations primarily focused on implementing structural engineering solutions to address riverbank erosion (Haque and Zaman 1994). An analysis of flood management and mitigation strategies in Bangladesh, focusing on the country's vulnerability to riverine risks. From the study, the institutional response from national and local governments, NGOs, and political parties, predominantly focus on providing food, and medication. Participants in the study areas reported receiving a small quantity of food aid and microloans. Out of our total of 66 participants, only 9 individuals stated that they received immediate assistance from either government or NGOs. This aid is consisted of tiny quantities of consumable items such as rice, dry food, and medicine, and often include small amounts of cash.

Union Councils are responsible for distributing the emergency assistance granted by the government. The governing body of these "Union Parishad" councils is composed of a Chair and twelve Members, with one Member specifically tasked with the distribution of aid within the study areas. The notification of relief is ineffective, and lacks defined guidelines. Usually, alerts are sent out in a central location within the village, such as a gathering spot or the local Union compound. This informal distribution of information often left out people who required help. One participant reported that relief supplies were given to his community without prior notification, resulting in individuals who were not present being excluded. Relief goods were often distributed based on the principle of prioritizing those who arrived first. In a previous research, Zaber et al. (2018) identified a prevailing sense of corrupt practices in the procedures related to relief and rehabilitation projects.

## 5.6 Community Response Addressing Riverbank Erosion

The community's response to riverbank erosion consisted of building community-based and partnerships within their personal social networks, such as friends, family, neighbors, and even coworkers. These collaborations were depended upon to effectively manage the responses, especially in cases when temporary or permanent migration was necessary. Prior to the flooding resulting from riverbank erosion, the individuals involved in the study area usually collect their belongings, such as furniture, utensils electronic devices, and construction equipment, and relocate to a temporary site. Over time, they relocated to farther areas that were less prone to riverbank eroippsion in order to find better economic prospects. The motivations for collaboration included a feeling of belonging to a community. According to the participants, when faced with flash flood or loss of land caused by erosion, they had to quickly find a safe location to construct shelters. The ancestral homes may be on their own land, or someone else's land through rent-based, or on government-owned land. However, landowners and impacted families sometimes formed informal agreements through their social connections. Given this situation, it is imperative to establish effective policies to assist individuals affected by river erosion, to prevent river bank erosion from intensifying the suffering of this vulnerable population in the long run.

## 5.7 The Nexus Between Riverbank Erosion, Migration, and Governance

In Bangladesh, riverbank erosion victims receive less attention from authorities compared to other kinds of natural disasters, such as tornado, and cyclone, who have higher priority on the list of disasters. Because riverbank erosion happens slowly and irregularly, the people affected by it are not able to get the attention of the authorities responsible for addressing the issue (Rana and Nessa 2017).

Similarly, riverbank erosion and displacement are not given media coverage like the victims of other natural hazards. Local people along the river in the study areas experience a nearly a limited response during the entire year. Further, victims of riverbank erosion are subject to differential treatment compared to victims of other natural catastrophes, and there is no dedicated policy or program in place from either the government or non-government sectors to address the issue of migrants displaced by riverbank erosion. Further, the government has released directives for the relief distribution to the affected due to disasters, resulting in unequal potential for displaced people due to the limited eligibility criteria for receiving such aid. Furthermore, the process of choosing victim families is also troublesome.

Victims of riverbank erosion lack temporary shelter, an early warning system, and an early evacuation strategy for horizontal erosion zones. Further, affected people forced to live below the threshold of extreme poverty as they grow more vulnerable in all aspects. To aid the individuals affected by riverbank erosion, it is necessary to revise the existing policies and actions of organizations. Riverbank erosion should be regarded as a separate and significant environmental disaster (Rana and Nessa 2017).

The disaster victims have been suffering due to the insufficient reaction from institutions in terms of formulating and implementing an adaptation strategy (Elahi and Rogge 1991; Rahman et al. 2015). Riverbank erosion, while distinct, can be reinforced with other disasters. Those affected by this phenomenon are provided assistance in line with disaster management rules and programs. The MWR and the MODMR are two significant ministries engaged in this domain. The MODMR is tasked with managing the government's disaster management initiatives in collaboration with the Directorate of Relief and Rehabilitation (DRR) and the Disaster Management Bureau (DMB). On the other hand, MWR is accountable for protection and management of riverbanks. The National Policy on Disaster Management, in general, has placed significant emphasis on community participation, and non-structural approaches to disaster risk mitigation.

There are three national-level bodies that focus on coordination across multiple sectors in Bangladesh. These bodies, as identified by Marzina Begum and Nurul

Momen (2019), are as follows: (a) The NDMC, which is led by the Prime Minister; (b) the IMDMCC, which is chaired by the cabinet minister responsible for the MODMR; and (c) the NDMAC, which consists of representatives from both the public and private sector. Additionally, there are other committees operating at the district, Upazila, and union levels, which are responsible for implementing the directions of the national committees.

The MODMR has the responsibility of increasing awareness before a crisis occurs, distributing supplies for relief after a disaster, and reconstructing the affected areas. The Khas land, Adarshaya Gram, and Abashon projects provide accommodation for certain groups of displaced individuals. NGOs, such as CARE-Bangladesh, OXFAM, BDPC, and RDRS collaborate with displaced individuals along the rivers in Bangladesh.

The national strategy on IDPs has been endorsed by the MODMR in order to facilitate the participation of displaced individuals in the development process. The strategy aligns with the objectives of the Delta Plan 2100 and the SDGs (MODMR 2021). Further, the GOB has formulated the National Strategy on the Management of Disaster and Climate Induced Internal Displacement (NSMDCIID) as a response to the Cancun Adaptation Framework, adopted in 2010 and the Sendai Framework for Disaster Risk Reduction (2015–2030). This strategy specifically focuses on addressing the needs of individuals who have been internally displaced due to natural disasters and climate-related hazards.

When it is not possible to prevent a disaster and to reduce the chances of people being compelled to leave their ancestral homes, it is important to establish strategies that protect individuals during evacuation and throughout their displacement (Adger et al. 2005). These measures should be based on rights and should continue until sustainable solutions such as returning to their original homes, integrating into local communities, or resettling elsewhere can be arranged. Additional measures are necessary from both the government and the international community.

The GOB's objective is to eradicate extreme poverty by 2030 and achieve upper-middle-income status by 2041, thereby ensuring the socioeconomic empowerment of the population. The strategic vision encompasses disaster risk management and climate change mitigation as integral parts. The government has been formulating and coordinating many social policies, programs, and strategies in line with its Social Development Framework (SDF) to achieve this goal. This paradigm aims to support Bangladesh in achieving social justice and equity in its development efforts, specifically by including migratory populations.

Building on the Sendai Framework for DRR, the government of Bangladesh (GOB) has transitioned from traditional response and relief strategies to a comprehensive risk reduction model. The objective of this shift is to strengthen the ability of communities to withstand the effects of climate change and disasters by aligning development and humanitarian activities with measures that specifically target reducing the risks associated with disasters (IPCC 2014).

The rights-based approach (RBA) sets crucial standards in its endeavor to provide lasting solutions for displacement, interventions to reduce risk, and humanitarian relief. The fundamental principles of the RBA to disaster and climate-induced

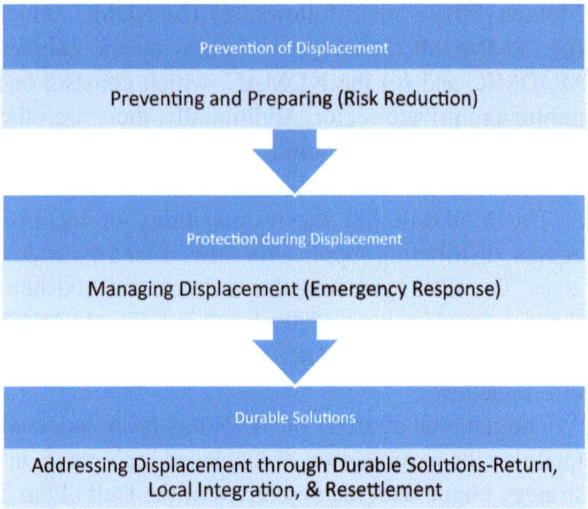


**Fig. 5.1** Displacement management framework (MODMR 2021)

internal displacement (DCIID) are based on the rights and privileges of the affected livelihood, which are acknowledged and protected by both national and international human rights agreements. The entitlements and rights also encompasses various socio-economic elements, along with the right to information and participate in decision-making processes (Momen and Ferdous 2023). In addition, the RBA places a high importance on addressing the needs and safeguarding the rights of marginalized groups, such as individuals living in poverty, those without a permanent home, people with disabilities, racial and religious minorities, and other vulnerable populations. This focus is justified by the fact that these groups are disproportionately exposed to risks in any situation involving internal displacement.

There is an approach that has created a DMF based on the IOM's Migration Management Cycle (MMC) and its integrated approach. This framework helps to determine suitable actions and interventions throughout different stages of displacement. The DMF is a practical and comprehensive approach since it categorizes several phases of displacement and develops strategic measures to tackle them. The human rights concerns of the displaced population have a considerable impact on the strategic actions adopted throughout the different stages of displacement (Islam and Rahman 2015). Their main goal is to safeguard the rights and privileges of the displaced individuals or households, with a particular emphasis on disadvantaged and marginalized groups. The guiding principles on IDPs also highlight the various obligations of states in relation to the prevention of displacement, the protection of individuals during displacement, and the facilitation of long-lasting solutions. Figure 5.1 illustrates the dimensions of the specific framework.

The above human rights-centered approach to the issue of displacement is founded on this principle. It highlights that it allows for both temporary and permanent displaced people and underscores the responsibilities of the state well before a specific hazard is about to happen. The graph illustrates that strategic interventions and responses during displacement are intended to minimize the impact of short-term displacement, whereas those carried out before disaster primarily aim to decrease the possibility of long-term displacement. On the other hand, interventions after displacement can also be effective and durable solutions to protect displaced individuals. The Strategy outlines four strategic approaches to the issue of DMF: Prevention, Preparedness, Management, and Addressing (MODMR 2021).

The main obstacles cover limited resources, insufficient oversight, climate change, and uncontrolled urban growth (Begum and Momen 2021). Nevertheless, there is currently no comprehensive state policy in place to effectively address displacement challenges due to river bank erosion. Some sectoral rules provide just an indication for rehabilitation strategies for displaced individuals, rather than clearly specifying full protective measures. In May 2007, the MODMR released four significant government circulars that provided distribution guidelines for CI Sheets, general relief, cash allotment for house building, and general relief for food to individuals affected by cyclones, fires, floods, tidal waves, and earthquakes (MODMR 2021). Affected by riverbank erosion are ineligible for the initial two categories of aid due to the required conditions. In order to be eligible for CI Sheets, individuals must possess land that they own. As a result, they are unable to acquire the required CI Sheets for building a temporary shelter. In the event of an individual's death resulting from a disaster, his or her family may be eligible to receive financial aid. Fortunately, the incidence of deaths due to riverbank erosion is infrequent. From the past experience in the study area, it is found that as a means of providing assistance, every affected household is given a 20 kg of rice, which is sufficient to sustain a family of five for a maximum of 10 days. Every family is allocated a sum of three thousand Taka to sustain their living, however, the process of selecting victim households is also a problematic. Although the government has built temporary shelters in places prone to cyclones, providing a safe place for residents to seek refuge during cyclone surges. Additionally, they can also make contact with those who do not visit the shelter because they live on their own homes. This also applies to individuals who have been affected by river bank erosion. Regrettably, victims of riverbank erosion lack access to any temporary shelter, as well as the absence of an early warning system and an arrangement for early evacuation for residents residing in areas prone to erosion. Due to the lack of potential for early alluvial deposition on their degraded property, individuals voluntarily leave the area and relocate to distant, safer locations after losing their home. The field study suggests that most of them move to a Sirajganj district (center). As a result, the local government and administrative authorities from the displaced people's original areas are unable to find and help them, on the other hand, the officials from the resettled administrative area have little capacity to support them, since they are not identified people affected of that administrative area. Consequently, these individuals who migrate due to river bank erosion become increasingly susceptible and are compelled to live below the poverty threshold. Further, in the earlier research has revealed that individuals affected by highly publicized disasters, such as the Bay of Bengal cyclones, for instance, Sidr, Bizli, Aila, etc., experience assistance when there is a coordinated effort to provide support and assistance from both private and public sectors. Conversely, a large number of individuals affected by river bank erosion, for example, coastal/riverbank erosion and sea level rise, are displaced due to natural disasters. These victims receive limited support to manage and adjust to their vulnerable situation and are not acknowledged as refugees, thereby missing out on the corresponding advantages like other natural disaster.

Developed countries have a crucial role in this situation by offering financial and logistical assistance. The increasing number of internally displaced individuals poses a challenge not only for the country experiencing the issue, but it also has the potential to result in 'inevitable migration'. To aid the individuals affected by riverbank erosion, it is necessary to revise the existing policies and actions of organizations. However, riverbank erosion is a severe environmental catastrophe that not only changes the physical landscape by encroaching on riverbanks but also leads to an increase in the number of IDPs who are compelled to live in slums.

During the river bank erosion, mobile money transfers could present difficulties in light of some conditions, such as extensive displacement of afflicted households to remote areas, limited phone charging capabilities, and complications in mobile money transfer (MMT. The analysis indicates that the communication and response of response organizations are inadequate, resulting in duplicated efforts and the negligence of the concerned authorities because of incorrect or arbitrary beneficiary selection (affected households). The interaction of disaster aid and sustainable livelihood reveals significant disparities between the immediate assistance given to internally displaced persons (IDPs), such as housing and food, and the support or aid aimed at their long-term recovery and development, including rehabilitation and recovery (Lobell et al. 2008).

## 5.8 Coping with the Crisis

At present, Bangladesh holds the seventeenth position with the highest number of IDPs in the world (BCCSAP 2009). IDPs in Bangladesh have encountered a wide range of encounters due to their socioeconomic and geographical causes, as well as the diverse events that have happened. However, vulnerability can be reduced by taking measures to reduce the loses of land and assets. Resilience, which stands in contrast to vulnerability, is a measure that assesses the propensity to prevent or minimize losses (Alexander 2002).

The affected individuals employ their indigenous wisdom and tactics as a means of managing the situation. As soon as the rainy season arrives and erosion begins to occur, they begin to relocate their possessions to secure locations. Initially, households affected with riverbank erosion evacuate their female members and children to safer locations. In addition, borrowing money at a high interest rate is the coping mechanism most frequently identified in the aftermath of displacement. Numerous individuals turn to selling their remaining productive assets, including livestock. The majority of individuals are unable to reconstruct their homes for lack of funds and space. As a result, they relocate and sell their equipment with significantly less money. The affected individuals choose to temporarily establish new homes in the

**Image 5.1** Areas near to the river *Location* Hatpachil, Koizuri union, Sirajganj District [*Photo credit* The author]

empty fields and streets or stay with relatives. Image 5.1 shows an area where the effects of riverbank erosion is visible.

The following Image 5.2 depicts the destructive force of Jamuna River in the study area.

When circumstances prevent children from attending school, they may be compelled to seek early employment (i.e. child labor) to supplement the households' income.

**Case story 1**

Md. Nazrul Islam, Boatman, Hatpachil, Koizuri, Shahajadpur

He is used stay near to Jamuna river bank the whole year and refers to the issue from his 20 years' experience. The tunnel are now formed with support of dredger. He suggests, at upstream layer of river, to lift sand and make the whole. It is important to refill soil equally, beside this tunnel. This type of tunnel may reduce the erosion, the authority should work from the deep of river.

The majority of respondents believe that the government and other stakeholders could enable IDPs to remain safely, as well as in the areas where they are displaced. Further, respondents believe that massive infrastructure development is necessary to prevent future catastrophes and build economic and social resilience. Diverse opinions exist regarding the accountability of government and other organizations. In

**Image 5.2** Silent cruelty of riverbank erosion *Location* Hatpachil, Koizuri Union, Sirajganj District [*Photo credit* The author]

order to address the concerns of IDPs, an interactive communication mechanism is essential in order to securely record complaints or offer feedback regarding the government, response organizations, and other entities concerned.

## 5.9 FGD: Perception from Respondents

The comprehensive riverbank erosion management policy should be adopted by the NDMC. The comprehensive policy is to assure the rights and safety of displaced persons in order to protect bank erosion and livelihood development programs (Begum and Momen 2023). Based on the preceding discussion, the following suggestions are made, according to the perceptions of the respondents, to reduce the vulnerability and increase the resilience capacity of those displaced by riverbank erosion. These programs should incorporate the below three phases of river bank erosion management (Table 5.2).

From the respondents, this research also found some other supportive guidelines for sustainable livelihoods for the affected people due to river bank erosion.

This may require an effective local government for preparedness for river bank erosion and involved non-governmental organizations, local civil society, and the private sector in the process. However, the respondents also specified that they lacked the capacity to mitigate riverbank erosion due to a lack of coordination between the concerned organizations.

**Table 5.2** Phase-wise initiatives and action based on data analysis

| Phase 1<br>Pre-displacement period | Phase 2<br>Response to emergencies | Phase 3<br>Post-relocation |
|---|---|---|
| – Identifying, evaluating, and monitoring potential river bank erosion risks and vulnerabilities<br>– The introduction of early warning system about river bank erosion. This ensures the prompt dissemination of early warning message<br>– Constructing storage systems at the district level to efficiently distribute relief and rehabilitation support<br>– Identifying khas land adjacent to secure areas in order to enhance the authority's capacity for management | – Perform the necessary tasks of rescuing, evacuating, and recovering individuals as required<br>– Providing temporary sanctuary and facilitating transportation may assist displaced individuals in reaching the shelter location<br>– Organizing evacuation zones and providing relief and medical items to individuals affected by river ban erosion<br>– Participating in rehabilitation services<br>– Improving recovery efforts, according to the requirements of the community | – Provide relief and recovery services<br>– Allocation of Khas land and the provision of housing, such as the model of Adarshya Gram and Abashon initiatives<br>– Implementing a banking policy that provides victims with interest-free loans and forgives their agricultural loans. Insurance plans with affordable premiums<br>– Establishing the necessary parameters (such as the depth of water, water flow velocities, and level of water) for different design alternatives (such as spurs and revetments) depending on the explicit attributes of main rivers<br>– Installing twin and triple-stages river bank protective measures |

It is important to develop a road map for the development of strategies for the river bank erosion for various entities. A detailed disaster risk management policy needs to be implemented to encounter the siltation along the river bed, which causes its profundity to diminish and its banks to erode. Further, sustainable embankment construction and its proper maintenance, massive afforestation with indigenous/local knowledge experience and action against deforestation may work against river bank erosion.

The relief and rehabilitation program must be implemented promptly and effectively. Many river erosion-affected individuals live below the poverty line and endure meager income, therefore, afflicted individuals should be offered low-interest loans. Priority should be given to the long-term programs. The necessity for long-term recovery programming is emphasized by the participants during the interview.

Social/public consciousness should be increased, especially after displacement. For instance, students dropping out of formal education, illiteracy is increasing in the area. Therefore, the government should take appropriate measures to manage it.

## 5.10 Conclusion

The socioeconomic disadvantages and marginalization of the residents along the river of the study areas add to the risks and vulnerabilities of their lives. The chapter highlighted its concerns about displacement due to river bank erosion in Sirajganj district in Bangladesh, and it underlines the significance of tackling the issue. This chapter emphasizes displaced households who must have access to secure shelter is a fundamental right. Furthermore, an understanding of households' risks and vulnerabilities and response strategies would aid policymakers in concentrating local adaptation and mitigation efforts, which are considered essential for enhancing the resilience capacity of the affected livelihoods and reducing vulnerability along the river in Bangladesh.

## References

Adger WN, Arnell NW, Tompkins EL (2005) Successful adaptation to climate change across scales. Glob Environ Chang 15(2):77–86. https://doi.org/10.1016/j.gloenvcha.2004.12.005

Alexander D (2002) Principles of emergency planning and management. Oxford University Press, New York

BCCSAP (2009) Bangladesh climate change strategy and action plan. Ministry of Environment and Forest, Dhaka

Begum MM, Momen MN (2021) Global climate change and its response. In: Maggino F (ed) Encyclopedia of quality of life and well-being research. Springer, Cham. https://doi.org/10.1007/978-3-319-69909-7_104663-1

Begum MM, Momen MN (2023) Legal frameworks of interstate disaster management. In: Romaniuk SN, Catino MS, Martin CA (eds) The handbook of homeland security, 1st edn. Taylor & Francis Group (Imprint: CRC Press), Boca Raton

Elahi KM, Rogge RJ (1991) Riverbank erosion, flood and population, displacements in Bangladesh: In: Elahi KM, Ahmed KS, Mafizuddin M (eds) Riverbank erosion, flood hazard and population displacement in Bangladesh. Riverbank Erosion Impact Study (REIS), Dhaka, p 364

IPCC (2014) Climate change 2014: impacts, adaptation and vulnerability, Part A: global and sectoral aspects, contribution of working group II to the fifth assessment report of the intergovernmental panel on climate change. Cambridge University Press, Cambridge. https://www.ipcc.ch/site/assets/uploads/2018/02/WGIIAR5-FrontMatterA_FINAL.pdf

Islam S, Rahman SH (2015) Coping with flood and riverbank erosion caused by climate change using livelihood resources: a case study of Bangladesh. Clim Dev 7(2):185–191. https://doi.org/10.1080/17565529.2014.910163

Lobell DB et al (2008) Prioritizing climate change adaptation needs for food security in 2030. Science 319:607–610. https://doi.org/10.1126/science.1152339

Marzina Begum M, Nurul Momen M (2019) Coordination does matter for disaster management in Bangladesh. In: Zutshi B, Ahmad A, Srungarapati AB (eds) Disaster risk reduction. Palgrave Macmillan, Singapore. https://doi.org/10.1007/978-981-10-8845-2_2

Mehedi H, Mukta Z, Roy K, Nag A, Farhana S (2010) Climate induced displacement: case study of cyclone Aila in the Southwest Coastal Region of Bangladesh. https://doi.org/10.13140/RG.2.1.3490.9921

MODMR (2021) Trend and impact analysis of internal displacement due to the impacts of disaster and climate change. Ministry of Disaster Management and Relief, Dhaka.

https://www.undp.org/bangladesh/publications/trend-and-impact-analysis-internal-displacement-due-impacts-disaster-and-climate-change
Momen MN, Ferdous J (2023a) Innovation and service delivery: conceptual and contextual framework. In: Governance in Bangladesh. New frontiers in regional science: Asian perspectives, vol 68. Springer, Singapore. https://doi.org/10.1007/978-981-99-0424-2_1
Momen MN, Ferdous J (2023b) Innovations in people's participation: mechanisms, practices, and challenges. In: Governance in Bangladesh. New frontiers in regional science: Asian perspectives, vol 68. Springer, Singapore. https://doi.org/10.1007/978-981-99-0424-2_7
Momen MN, Baikady R, Sheng-Li C, Basavaraj M (2020) Introduction: civil society in the era of globalization—emerging concerns as a social welfare provider. In: Momen MN, Baikady R, Sheng Li C, Basavaraj M (eds) Building sustainable communities. Palgrave Macmillan, Singapore. https://doi.org/10.1007/978-981-15-2393-9_1
Rabbi H, Saifullah ASM, Sheikh MS, Sarker MMH, Bhowmick AC (2013) Recent study on river bank erosion and its impacts on land displaced people in Sirajgonj Riverine Area of Bangladesh. World J Appl Environ Chem Int J Appl Environ Sci 2(2):36–43
Rahman MATMT, Islam S, Rahman SH (2015) Coping with flood and riverbank erosion caused by climate change using livelihood resources: a case study of Bangladesh. Clim Dev 7(2):185–191. https://doi.org/10.1080/17565529.2014.910163
Rana MS, Nessa AM (2017) Impact of riverbank erosion on population migration and resettlement of Bangladesh. Sci J Appl Math Stat 5(2):60–69. https://doi.org/10.11648/j.sjams.20170502.11
Shamsuddoha M, Chowdhury RK (2007) Climate change impact and disaster vulnerabilities in the coastal areas of Bangladesh. COAST Trust and Equity and Justice Working Group, Dhaka
Walsham M (2010) Assessing the evidence: environment, climate change and migration in Bangladesh. International Organization for Migration (IOM). https://publications.iom.int/system/files/pdf/environment_climate_change_bangladesh.pdf
Zaber M, Nardi B, Chen J (2018) Responding to riverbank erosion in Bangladesh. In: COMPASS'18: ACM SIGCAS conference on computing and sustainable societies (COMPASS), June 20–22, 2018, Menlo Park and San Jose, CA, USA. ACM, New York, NY, USA, 11 p. https://doi.org/10.1145/3209811.3209823

# Chapter 6
# Conclusion: Riverbank Erosion Vulnerabilities and Its Coping Strategies in Bangladesh

**Abstract** The concluding chapter highlights the main findings of the study and addressing the deficiencies in the current support programs for the affected households due to river bank erosion in Sirajganj district. This study has revealed the complex effects of riverbank erosion on the livelihoods of communities in Bangladesh, where rivers significantly influence both the landscape and the socio-economic structure. This study explore the close connection between environmental change and social vulnerability, demonstrating how erosion impacts individuals and communities variably according to their socio-economic status, gender, and geographical location. In Bangladesh, it is also crucial to include the marginalized voices of those affected by river bank erosion in governance processes. This may be achieved by their active involvement and capacity building. By adopting a bottom-up approach, Bangladesh can effectively address the SDGs. However, the preceding chapter has drawn the findings from a deliberately selected interview with 66 individuals from 54 families, including 20 households from urban slums in Dhaka, FGDs, KIIS, and case studies. Detailed justification of data collection is provided in Chap. 1. The findings show that the displaced population due to erosion along the river receives no significant support in securing livelihood, especially in reconstructing homes and/or opportunities in income generation.

## 6.1 Introduction

This study supports the idea, to a large extent, as suggested by Hunter et al. (2015), that it is worth to take a context-specific research when studying the complex vulnerability of IDPs due to river bank erosion. At some point the factors that contribute to vulnerability can present themselves as contextual problems in a more dynamic way, taking into account the combined impact of multiple factors in a disaster situation (Momen et al. 2024). This study highlights the importance of providing better assistance to riverine areas where vulnerable displaced and their families live. Additionally, it clarifies that when adult family members engage in

 
M. Begum, *Understanding River Bank Erosion in Bangladesh*, Springer Water,
https://doi.org/10.1007/978-3-032-00288-4_6

short-term migration, it may reduce the vulnerability experienced by the family members who remain in their original location.

## 6.2 Impacts on Local Livelihood Resulting from River Bank Erosion

Riverbank erosion has resulted in considerable displacement, leading several people to migrate in pursuit of safety and improved opportunities. This frequently leads to congestion in Dhaka when migrants encounter difficulties such as insufficient housing, restricted employment opportunities, and social alienation though there are few exceptions also. In the short term, forced migrants face urgent obstacles include securing shelter, obtaining clean water, and accessing employment prospects. These initial shocks impacted families into a condition of socio-economic instability. Ultimately, these displacements lead to significant transformations, including modifications in community structures, the erosion of conventional social networks, and a rise in urban slum populations. The research has discovered that the immediate consequences entail displacement and loss of livelihood, whereas the prolonged repercussions encompass a transition towards urbanization, as displaced individuals relocate to cities in pursuit of enhanced chances.

Zaman also contends that migration resulting from erosion causes a progressive deterioration of rural economies and exacerbates poverty in metropolitan regions, as unskilled laborers swamp the employment market. In another study Paul and Routray (2013) argued Forced migration, exemplified by congested urban slums, heightened employment competitiveness, and the breakdown of traditional livelihood practices. The narrative of rural-urban migration, while significant, may oversimplify intricate socio-economic considerations, including the reasons some individuals prefer not to relocate despite the associated risks, such as strong cultural connections to their land. Alam et al. (2018) argue that non-migration, frequently disregarded, significantly influences the comprehension of vulnerability dynamics.

## 6.3 Assessing the Patterns and Effects of Livelihood Vulnerabilities

The primary aim of the study was to investigate the various patterns and effects of livelihood risks and vulnerabilities within the impacted people in Bangladesh. Local communities impacted by erosion have been compelled to diversify their livelihoods as conventional agriculture practices become unfeasible. While some individuals have secured alternate work, others have encountered difficulties in adapting, underscoring inequities in access to services and knowledge. The patterns of vulnerability exhibited considerable variation influenced by socio-economic characteristics, including income level, education, and asset ownership. Lower-income demographics saw more profound repercussions, encompassing the loss of land, housing, and employment. Previous researchers, including Islam and Rashid (2011) and Rahman and Gain (2020) have also established that riverbank erosion constitutes both an environmental threat and a socio-economic issue. Again, research conducted by Hugo (2011) indicates that migrants, despite initial vulnerability, frequently acquire new skills and tactics that improve their adaptation to changing environments.

### 6.3.1 Economic Vulnerabilities

The most apparent and tangible expression of vulnerability is economic in nature. Families lose their primary sources of income when their land and property get submerged by rivers. Agriculture serves as the primary source of subsistence for rural communities, and the depletion of arable land compels individuals into a state of landlessness. In the absence of alternative income streams, displaced families endure constant poverty. Islam and Rashid (2011) emphasize that the degradation of agricultural land from erosion adversely impacts food security and household income, hence exacerbating poverty for numerous families. Moreover, landless families encounter challenges in obtaining credit or loans due to their absence of the collateral generally demanded by financial institutions. Consequently, they frequently resort to informal, low-paying employment, which is unstable and inadequate to satisfy their fundamental need. In addition, Research conducted by Paul and Routray (2013) indicates that relocated populations are more susceptible to trap in cycles of poverty owing to restricted options for growth in urban settings. Moreover, their deficiency in education, skills, and social capital inhibits their access to more stable and higher-paying employment, resulting in prolonged economic vulnerability following displacement.

### 6.3.2 Social Vulnerabilities

Social vulnerabilities relate to the disproportionate impact of riverbank erosion on various population segments, especially marginalized and vulnerable populations. Women, specifically, encounter considerable obstacles in this situation. In numerous rural regions, women predominantly manage domestic tasks and rely on subsistence agriculture or handicrafts to augment the family income. Erosion results in the loss of income sources and imposes more costs in family management. Adger (2006) also argues that social vulnerabilities are intimately linked to social norms, power dynamics, and resource accessibility. Women frequently possess diminished influence over familial resources, including land and finances, hence complicating their ability to recover from displacement and economic adversity.

Furthermore, the collapse of social networks resulting from displacement exacerbates the isolation of vulnerable populations, constraining their capacity to seek assistance or engage in communal decision-making. Rayhan and Grote explore that the disintegration of social connections might hinder displaced individuals in reconstructing their lives, as they lose access to informal networks that once assisted them in securing employment, sharing resources, or offering emotional support. In a study Rahman and Gain (2020) establishes an intergenerational cycle of marginalization, wherein the enduring results of forced migration persistently impact next generations.

Unplanned urbanization, stemming from rural-to-urban migration caused by riverbank erosion, has led to the emergence of informal settlements, where displaced families reside in hazardous and unsanitary conditions. This has also led to an increase in urban poverty, as numerous migrants struggle to secure steady employment in the formal sector.

### 6.3.3 Psychological Vulnerabilities

Psychological vulnerabilities are frequently neglected. Psychological vulnerability is associated with social isolation, as displaced individuals frequently encounter difficulties in adapting into new communities. The pain of losing one's home, land, and means of subsistence has lifelong emotional consequences for those impacted. Displaced families, particularly children, may encounter mental health challenges including anxiety, despair, and feelings of hopelessness. The psychological stressors are exacerbated by future uncertainty, since displaced families often relocate numerous times in search of stability. In a research Das et al. (2014) indicates that recurrent relocation caused by erosion results in "chronic vulnerability," preventing affected individuals from fully recovering from one disaster before encountering another. This ongoing cycle of relocation engenders a continual sense of fear and emotional distress.

### 6.3.4 Gendered Vulnerabilities

Gender significantly influences the patterns of vulnerability associated with riverbank erosion. Women and girls are disproportionately impacted, encountering distinct issues such as heightened vulnerability to gender-based violence, diminished access to healthcare, and a decline in educational possibilities. Displaced households are more likely to withdraw girls from school to assist with domestic responsibilities or engage in wage jobs to sustain the home. Rahman and Gain (2020) indicates that gender inequities intensify after displacement, with women and girls encountering increased obstacles in resource access and decision-making participation. They are additionally more susceptible to exploitation and abuse in temporary shelters or urban slums where displaced families frequently reside. This establishes a cycle in which gendered vulnerabilities both originate from and increase the wider socio-economic consequences of riverbank erosion.

### *6.3.5 Environmental Vulnerabilities*

Over time, displacement exerts environmental effects, as displaced populations contribute to the deterioration of environment in their new areas. The expansion of informal settlements in cities frequently results in pollution, and the degradation of environment. On the other hand, in rural regions, the relocation of displaced populations may result in the overexploitation of land and water resources, intensifying environmental degradation in already vulnerable ecosystems. Mutton and Haque (2004) also stated that the displacement of individuals resulting from riverbank erosion is integral to a broader cycle of environmental degradation, wherein the displacement compounds further environmental harm.

## 6.4 Administrative Competence and Governmental Assistance

Another important objective of the study aimed to assess the administrative capabilities and governmental responsiveness to the consequences of riverbank erosion. The research indicated that despite the Bangladeshi government's initiatives to mitigate riverbank erosion through embankment construction and financial compensation, these measures frequently prove insufficient due to limited financing, and poor coordination. Furthermore, disaster management frameworks in the nation often emphasize emergency aid rather than long-term resilience development (Adger 2000). It is noted that local administrative authorities frequently lack resources, hindering their ability to properly address the needs of displaced communities. Furthermore, there is an absence of comprehensive strategies that emphasize rehabilitation and sustained support for the impacted populations. This constraint prolongs a persistent cycle of vulnerability, since impacted populations receive little aid in reconstructing their lives.

Riverbank erosion in Bangladesh constitutes an extensive and persistent natural disaster, impacting thousands of individuals each year in Bangladesh. The government's ability to manage this threat, deliver prompt assistance, and secure sustainable recovery for displaced communities is vital. The study on patterns and livelihood vulnerabilities resulting from riverbank erosion identifies significant deficiencies in administrative capability and governmental assistance. The issues arise from constrained resources, inadequate long-term planning, bureaucratic inefficiencies, and inadequate institutional coordination.

### 6.4.1 Sustainable Housing Solutions

A notable deficiency in governmental assistance is the absence of enduring housing alternatives for displaced families. Individuals displaced by riverbank erosion often resort to temporary shelters, improvised dwellings on embankments, or urban slums. The conditions are unclean and hazardous, rendering displaced families susceptible to more displacement or health risks. Although certain government initiatives have sought to offer land or houses to displaced families, the execution of these programs frequently exhibits deficiencies. Corruption in SSN, bureaucratic obstacles, and inadequate cooperation between national and local agencies sometimes obstruct displaced individuals from obtaining land or housing rights. Furthermore, the land designated for resettlement is frequently situated in environmentally vulnerable or geographically remote regions, so constraining the livelihood prospects for the displaced.

### 6.4.2 Land and Property Rights

A significant long-term challenge confronting displaced communities is the absence of land and property rights. Riverbank erosion frequently results in the permanent displacement of individuals. The government has undertaken efforts to allocate khas land (government-owned property) to the displaced; however, the process is frequently hindered by legal issues, and inefficiency. Land distribution initiatives in Bangladesh are hindered by inadequate governance, with powerful landowners or local elites frequently exploiting the system to appropriate land designated for the displaced. Consequently, numerous relocated families are either left landless or allocated land that is unsuitable for agriculture or other economic pursuits. The absence of secure land rights intensifies the vulnerability of displaced communities, placing them in a vulnerable position with few opportunities for recovery.

### 6.4.3 Khas Land Allocation and Management

Government initiatives to distribute khas land (state-owned property) to landless families have been integral to rehabilitation techniques; yet, these programs frequently have inefficiencies. Enhanced governance and monitoring systems are essential in ensuring equitable distribution of khas land to those in need most. It is noted that the government must ensure that land allocated to displaced people is appropriate for agricultural or other income-producing endeavors. Often, the designated land is situated in isolated regions, thereby restricting its utility. Enhancing the quality of allocated land and supplying essential resources for displaced households to utilize it efficiently is a vital measure for sustainable recovery.

### 6.4.4 Coordination Among Various Government Departments and Stakeholders

A significant obstacle to efficient disaster management and reconstruction in Bangladesh is the inadequate coordination among various government departments and stakeholders. However, riverbank erosion is a multifaceted problem that encompasses several sectors, including disaster management, water resource management, and social welfare. Nonetheless, the absence of cohesive planning and communication among these sectors frequently leads to scattered and inefficient responses.

### 6.4.5 Collaboration Between National and Local Agencies

The national government has established rules and procedures for disaster management, although their execution at the local level sometimes lacks efficacy. Local government entities, tasked with providing services and assistance to displaced persons, sometimes lack the competence, or resources to execute national policy efficiently. Choudhury et al. assert that the centralization of authority in disaster management results in a disjunction between the national government's strategies and the actual conditions experienced by local populations. The participation of excessive stakeholders without defined roles and responsibilities results in overlapping mandates, bureaucratic inefficiencies, and delays in aid delivery. Effective disaster management necessitates a more decentralized strategy, wherein local governments are endowed with the requisite resources and decision-making capacity to promptly address the needs of displaced populations.

### 6.4.6 Inconsistencies in Policy Performance

A significant difficulty is the inconsistency in policy execution. Although Bangladesh has established many disaster management policies and plans, such as the National Plan for Disaster Management (NPDM) and the Standing Orders on Disaster (SOD), their implementation is sometimes incomplete or uneven. The failure to implement policy promises results in deficiencies in support for displaced populations, especially in rehabilitation, social protection, and livelihood restoration. Islam and

Walkerden noted that the government's emphasis predominantly lies on rapid catastrophe response and relief, neglecting long-term adaptation and resilience-building initiatives. The riverbank erosion control from comprehensive development planning diminishes the government's capacity to offer adequate assistance to impacted communities.

## 6.5 Significant Research Findings: Lessons Learned

The findings suggests expanding targeted assistance programs (refer to the fifth chapter), such as income generating initiatives, access to institutional financing for those without land, more affordable transportation, and greater communication. The researcher finds that single factors do not independently contribute to a community's vulnerability to river bank erosion, instead, several factors interact and strengthen each other. Thus, this study suggests improving infrastructure by improving irrigation services, providing affordable transportation for agricultural products and materials, ensuring appropriate prices for everyday necessities, and enhancing health care advantages. Several investigations have shown that erosion leads to livelihood disruption, making it an inevitable and undeniable aspect of life for people living along river banks (Piguet 2013; Cosgrave 2014; Walsham 2010; Momen 2021). An increase in rainfall during the monsoons is believed to be a contributing factor to a higher erosion rate along the river in Bangladesh, which could result in a loss of residential and agricultural land (BCCSAP 2009).

Riverbank erosion poses a significant threat to the lives and means of subsistence of displaced individuals in multiple ways. This study has demonstrated the immediate and indirect detrimental consequences of internally displaced persons (IDPs), with forced displacement being the most devastating outcome. The displacement caused by riverbank erosion severely destroys infrastructure, environmental safety, thus impacting the livelihood pattern. In order to address these problems, it is highly encouraged that the GoB should promptly formulate an all-inclusive national strategy to safeguard IDPs. The term 'river refugee' are not officially recognized in the lexicon of forced migration in the legal and institutional framework in Bangladesh. This book proposes that the GoB should develop a comprehensive national strategy to safeguard internally displaced persons (IDPs), especially those who have been impacted by riverbank erosion.

River bank erosion is a persistent phenomenon that causes both direct and indirect damage. The results underscore the necessity for more comprehensive policies and initiatives that deliver urgent assistance while also promoting enduring resilience. Potential measures may cover the establishment of sustainable livelihood alternatives for displaced individuals, and investment in education and skills development for impacted communities. Future study may investigate the convergence of climate change adaptation measures and the government's role in fostering sustainable development, especially in areas susceptible to erosion. Furthermore, greater emphasis should be placed on the interaction between these displacement patterns and other social challenges. The subsequent sections outline critical policy implications and prospective strategies for mitigating livelihood vulnerabilities resulting from this major problem. In addition, the following recommendations are provided for addressing river bank erosion and supporting affected households.

### 6.5.1 Guiding Principles on Displacement

A critical policy issue is the necessity to formally incorporate displacement from riverbank erosion into national disaster management frameworks. Despite its considerable socio-economic consequences, displacement due to riverbank erosion has not received equivalent attention as other natural disasters, such as cyclones or floods, in national policy formulation.

Displacement should be seen as a last resort, and it frequently results in the marginalization of the most vulnerable people living in the riverbank areas. Landless families are frequently excluded from temporary shelter efforts. However, the national guiding principles on Internal Displacement should include the input of key stakeholders, particularly those affected by river bank erosion. It is imperative that we align norms pertaining to the safeguarding of internally displaced persons (IDPs). Additionally, with regards to the livelihoods and way of life of individuals displaced by riverbank erosion, a unique policy for their rehabilitation should include specific measures.

The National Plan for Disaster Management (NPDM) establishes a framework for managing diverse calamities in Bangladesh. Riverbank erosion is frequently regarded as a localized problem rather than a national issue. Explicitly incorporating riverbank erosion into the NPDM would guarantee the allocation of sufficient resources, research, and planning to address this threat. This inclusion will enhance coordinated efforts across government agencies, NGOs, and international organizations. Mutton and Haque (2004) assert that policies prioritizing riverbank erosion in conjunction with other risks will force the government to formulate comprehensive strategies to tackle both immediate and long-term impacts. Prioritizing riverbank erosion within the NPDM enables the government to increase early warning systems, improve disaster preparedness, and formulate targeted interventions for at-risk communities (Mutton and Haque 2004).

The Standing Orders on Disasters (SOD) offer directives for governmental bodies in disaster management; nevertheless, they insufficiently include the particular dynamics of riverbank erosion. Modifying the SOD to incorporate explicit rules for addressing riverbank erosion may elucidate the duties and obligations of different governmental entities, particularly regarding the coordination of relief initiatives and long-term rehabilitation activities. Further, incorporating riverbank erosion into current disaster management policies can mitigate the institutional fragmentation that presently obstructs relief and restoration initiatives.

### 6.5.2 Promoting Climate Change Adaptation

Climate change contributes to riverbank erosion, as rising sea levels and intensified river flows lead to more frequent and severe erosion occurrences. Consequently, tackling the problem of riverbank erosion necessitates the incorporation of climate change adaptation measures into both national and local policies. Hence, it is important to integrate climate change projections into land use and development planning. Climate change policies should promote adaptive practices, such as crop diversification and sustainable agricultural techniques, to enhance resilience. However, Bangladesh has formulated many climate change adaptation measures, including the Bangladesh Climate Change Strategy and Action Plan (BCCSAP). Nevertheless, these policies must be more closely integrated with disaster management methods to effectively address the particular risks linked to riverbank erosion. Future strategies must prioritize enhancing the adaptive capacity of at-risk communities via education, technology transfer, and financial assistance.

### 6.5.3 Significance of Sustainable Livelihoods

Everyone in each country aspires to have a stable and sustainable means of supporting themselves. To evaluate the livelihood challenges faced by the impacted individuals, implementing a robust strategy can promote the development of sustainable livelihoods for everyone. In order to develop a comprehensive policy to improve the lives of impacted households, it is crucial to identify and understand the vulnerable characteristics they experience. Nevertheless, it is imperative for the GOB to give priority to expeditiously restoring the means of generating a living expenses in areas that have been significantly impacted by river erosion. The concerned organization must promptly implement a proactive approach to identify essential initiatives for ensuring food security, shelter, and ultimately improving people's livelihoods. These initiatives should have the objective of restoring assets and generating income for the communities impacted by river bank erosion.

### 6.5.4 Enhancing Organizational Capability and Strengthening Local Governance Institutions

Governments and local institutions must collaborate to create legal safeguards for displaced communities that promote recovery and adaptation. The government's ability to respond successfully to riverbank erosion, relies on the operational efficacy of institutions such the Ministry of Disaster Management and Relief, local government entities, and the Bangladesh Water Development Board (BWDB). Nevertheless, these organizations frequently experience resource deficiencies, inadequate inter-agency coordination, and bureaucratic delays, which diminish their capacity to execute prompt and effective catastrophe responses. Haque and Zaman (1994) stated that institutional deficiencies and inadequate governance frameworks intensify the plight of displaced persons. Local government entities, intended to be the primary responders, frequently lack adequate resources, skilled staff, and equipment to fulfill their duties efficiently (Haque and Zaman 1994). Moreover, local administrative units sometimes possess overlapping tasks, resulting in confusion and inefficiencies in the allocation of help and resources.

Participation of the local government and the affected population is vital. Decentralizing disaster management and augmenting local governments with more decision-making authority and resources would facilitate improved catastrophe response times and ensure that relief effectively reaches impacted communities. Local governments are frequently hindered by insufficient finance and technical expertise, restricting their potential to execute national objectives. Haque and Zaman (1994) stated that local governments must enhance their engagement in the development and execution of disaster management policies. Decentralized governance will provide more adaptable and contextually relevant responses to riverbank erosion, as local authorities possess a superior understanding of their people' needs (Haque and Zaman 1994). However, the study on livelihood vulnerabilities reveals that local governments possess a limited capacity to manage riverbank erosion and its consequences effectively. Strengthening local governance is essential for the efficient implementation of disaster management strategies and for providing sufficient support to displaced populations.

### 6.5.5 Community Resilience

In spite of the significant vulnerabilities induced by riverbank erosion, impacted communities adopt adaptive strategies to adjust to their evolving circumstances. Families frequently depend on social networks for assistance, encompassing extended relatives or community organizations. In several instances, communities have embraced migration as a coping mechanism, with one or more family members relocating to urban centers or overseas to pursue employment and remit funds back home. Alam et al. (2018) characterize migration as an expression of "adaptive resilience," wherein displaced people or families move to mitigate their vulnerability to risks and obtain alternate means of subsistence. This adaptation pays a cost, frequently resulting in the breakup of family units and increasing dependence on uncertain low-wage jobs in urban areas (Alam et al. 2018).

This highlights the necessity for more comprehensive policies that offer sustainable alternatives to migration and tackle the root causes of vulnerability. It is also needed to emphasize resilience-enhancing interventions, including vocational training, education, and access to social services, in assisting displaced families to reconstruct their life and mitigate long-term vulnerability. Policies targeting the fundamental causes of displacement, including enhancing disaster management, offering social protection, and ensuring access to land and resources, are crucial for establishing long-term solutions.

Further, it is also need to allocate resources to community-driven activities that strengthen social capital and build local resilience. This comprises training programs that enable individuals to establish adaptive strategies. The capacity of communities to address erosion is significantly influenced by their social capital. Individuals with strong connections generally do better, highlighting the significance of community cohesion and collaborative efforts in alleviating vulnerabilities.

### 6.5.6 Community-Based Adaptation Strategies

It is important to engage local communities in developing and implementing erosion control measures. Community participation ensures that strategies are culturally relevant and more likely to succeed. Community-based disaster management (CBDM) strategies, which engage local people in decision-making and planning, have demonstrated enhanced catastrophe resilience. By engaging community people in the formulation and execution of disaster risk reduction plans, governments can ensure that interventions are more relevant and adapted to local requirements. Alam et al. (2018) also suggest that enabling local populations to engage in disaster management results in more sustainable outcomes. Further, active community involvement can provide local knowledge and resources, hence improving the efficacy of disaster preparedness and response strategies. Future policies must emphasize the establishment of community-driven disaster management programs to enhance local people' ability to address riverbank erosion and other environmental threats (Alam et al. 2018).

### 6.5.7 Infrastructure Resilient to Climate Change

It is important to develop robust infrastructure, such as riverbank protection works (e.g., embankments and revetments), to minimize erosion. Investments should also focus on improving access to markets and services for displaced communities. Investment in climate-resilient infrastructure, including river embankments, flood barriers, and erosion control systems, is crucial for mitigating the risk of displacement caused by riverbank erosion. Infrastructure projects must be tailored to the unique natural conditions of Bangladesh and should engage local communities in their development and execution. Paul and Routray (2013) stated that forthcoming strategies must emphasize investments in infrastructure that safeguard against erosion while fostering sustainable livelihoods. Integrating erosion control with irrigation systems can simultaneously offer erosion protection and agricultural assistance, thus fulfilling both immediate and enduring requirements.

### 6.5.8 Improving Public Awareness and Education

It is important to promote public awareness campaigns to educate communities about riverbank erosion and its implications. Empowering communities with knowledge can foster proactive adaptation measures. It is imperative to implement compulsory community education with a focus on raising awareness in order to decrease vulnerability in the long term. A clear and comprehensive guide will assist individuals who are vulnerable to effectively manage the impact of river bank erosion. Similarly, the utilization of various approaches, along with the application of indigenous knowledge, can effectively reduce vulnerability and foster the development of resilient communities.

### 6.5.9 Establishing Quality Networks

An essential requirement for reducing vulnerability is the establishment of a high-quality network connecting researchers, academicians, practitioners, and the general public. In addition, this network should strengthen the connection with various national accomplishments, especially during certain events. Riverbank erosion is a longstanding issue in the history of Bangladesh. Therefore, the government still has the opportunity to collaborate with NGOs and other international organizations to address the severe vulnerability caused by this problem.

### 6.5.10 Exploring Social Safety Net and Reconstruction Programs

Diverse social programs can enhance the abilities of individuals at all stages within the society. Within this program, individuals from all categories, such as adolescents, elderly, males, females, and even children, can be involved in order to attain inclusive community development. Conditional cash transfers for shelter can serve as an additional or substitute approach when catastrophic events result in substantial housing damage. Further, the Money for Work program is a very efficient strategy for focusing on disadvantaged households, as it has been designed for individuals who are underprivileged. When preparing a response for Money for Work, it is also important to consider that individuals with chronic illnesses and impairments may be unable to participate in this program. In order to effectively reconstruct and strengthen embankments, it is essential for Money for Work initiatives to incorporate adequate technical oversight and establish mutually agreed-upon standards with development partners.

To mitigate river bank erosion, it is necessary to construct an engineered embankment and strategically put sandbags. Moreover, the research results suggest that assessing the impact of riverbank erosion on the food security of individuals living in areas affected is an essential requirement for formulating effective policies for rural development. It is important to establish financial safety nets for affected communities, tailored for those vulnerable to erosion. This can help mitigate the economic impacts of displacement and loss of livelihoods (Reid and Huq 2007). Further, it is imperative to include nutrition-sensitive methods into the interventions offered to affected individuals. It is imperative to prioritize the continuous execution of sanitation and water services in order to effectively control the spread of diseases.

## 6.6 Policy Implications and Future Directions

To tackle river bank erosion, a thorough comprehension of the fundamental causes and dynamics of riverbank erosion is essential. Studies indicate that soil characteristics, human activities, and hydro geomorphological processes significantly contribute to river bank erosion and channel instability (Okeke and Ede 2019). Ongoing scientific research and monitoring of these variables can guide more effective governmental initiatives.

A possible method is the strategic dredging of riverbeds, which can enhance navigability and mitigate erosion risk. Nonetheless, the enduring effects of these dredging operations on other segments of the river system require meticulous assessment to guarantee sustainable management (Islam et al. 2018). Improving community resilience is an essential aspect of a comprehensive policy response. Providing impacted populations alternative livelihood opportunities, disaster preparedness education, and access to social safety nets can assist them in managing the economic challenges resulting from riverbank erosion. Policymakers must plan the for the consequences of climate change and its capacity to intensify riverbank erosion. Budgetary allocation in climate-resilient infrastructure, like embankments and flood control systems, can reduce the effects of extreme weather events and safeguard at-risk communities (Islam et al. 2018; Hossain 1993; Bhuiyan et al. 2017; Okeke and Ede 2019).

Tackling the complex issue of riverbank erosion in Bangladesh necessitates a comprehensive strategy that integrates scientific inquiry, targeted governmental measures, and community-oriented resilience initiatives.

However, the patterns of livelihood vulnerabilities resulting from riverbank erosion in Bangladesh highlight the necessity for effective policy responses. Effective disaster management and long-term rehabilitation necessitate urgent relief as well as comprehensive strategies that tackle the underlying causes of vulnerability, enhance resilience, and foster sustainable development. The research highlights the policy implications and future strategies required to alleviate the adverse effects of riverbank erosion and enhance the livelihoods of affected communities. Effective policy responses must incorporate local expertise and community involvement to provide lasting solutions. There is an urgent necessity for thorough catastrophe risk management strategies that take into account the socio-economic settings of impacted communities. However, continued research is vital for understanding the long-term impacts of riverbank erosion. This includes monitoring social and economic changes in the affected communities to inform future interventions.

## 6.7 Conclusion

It is worth to conclude by saying that while some factors contributing to risk and vulnerability, such as exposure to riverbank erosion and reliance on natural systems, have been explored in the context of bio-physical aspects. Further, the analysis suggests that economic and infrastructural inequalities play a significant role in generating various factors of risks and vulnerabilities. These drivers are further shaped by a diverse range of underlying narratives derived from the perception of the affected individuals. Thus, this research states that many of these drivers are interconnected and do not independently expose the riverine community to risks and vulnerabilities. The risks and vulnerabilities of the affected households subjected to riverbank erosion is not exclusively determined by high environmental susceptibility or other bio-physical issues or economic factors, instead, these disparities arise from a combination of interconnected factors, including economic, institutional, infrastructural, and geographical inequities, which are worsened by inadequate infrastructure and corrupt practices in rehabilitation and reconstruction.

In conclusion, tackling the difficulties of riverbank erosion in Bangladesh necessitates a comprehensive understanding of the interaction between environmental change and social dynamics. This research not only enhances scholarly discourse but also demands immediate action to save vulnerable populations from the uncompromising power of nature. By emphasizing community resilience, and inclusion, the society can more effectively assist impacted populations and promote sustainable development despite persistent river bank erosion difficulties.

Mitigating the vulnerabilities resulting from riverbank erosion in Bangladesh necessitates a comprehensive strategy that includes sustainable policy development, community involvement, and inventive solutions as well. By emphasizing integrated management measures and investing in community resilience, Bangladesh can alleviate the detrimental impacts of erosion and improve the livelihoods of the most impacted populations. The cooperation of government and local communities could be crucial for developing sustainable solutions to the river bank erosion problem.

## References

Adger WN (2000) Social and ecological resilience: are they related? Prog Hum Geogr 24(3):347–364. https://doi.org/10.1191/030913200701540465

Adger WN (2006) Vulnerability. Glob Environ Chang 16(3):268–281. https://doi.org/10.1016/j.gloenvcha.2006.02.006
Alam GMM, Alam K, Mushtaq S (2018) Climate change perceptions and local adaptation strategies of hazard-prone rural households in Bangladesh. Clim Risk Manag 17:52–63. https://doi.org/10.1016/j.crm.2017.06.006
BCCSAP (2009) Bangladesh climate change strategy and action plan. Ministry of Environment and Forest, Dhaka, Sept 2009
Bhuiyan MAH, Islam SMD-U, Azam G (2017) Exploring impacts and livelihood vulnerability of riverbank erosion hazard among rural household along the river Padma of Bangladesh. Environ Syst Res 6(25). https://doi.org/10.1186/s40068-017-0102-9
Cosgrave J (2014) Responding to flood disasters: learning from previous relief and recovery operations. ALNAP lessons paper. ALNAP/ODI, London. Available at https://library.alnap.org/system/files/content/resource/files/main/alnap-lessons-flood-cosgrave.pdf
Das KT, Halder HK, Gupta ID (2014) River bank erosion induced human displacement and its consequences. Living Rev Landscape Res 8(2014):3. https://doi.org/10.12942/lrlr-2014-3
Haque CE, Zaman MQ (1994) Vulnerability and responses to riverine hazards in Bangladesh: a critique of flood control and mitigation approaches. In: Disasters, development and the environment. Wiley, New York
Hossain M (1993) Economic effects of riverbank erosion: some evidence from Bangladesh. Disasters 17(1):25. https://doi.org/10.1111/j.1467-7717.1993.tb00485.x
Hugo G (2011) Future demographic change and its interactions with migration and climate change. Glob Environ Chang 21(1):521–533. https://doi.org/10.1016/j.gloenvcha.2011.09.008
Hunter LM, Luna JK, Norton RM (2015) Environmental dimensions of migration. Annu Rev Sociol 41:377–397. https://doi.org/10.1146/annurev-soc-073014-112223
Islam MF, Rashid AB (2011) Riverbank erosion displaces in Bangladesh: need for institutional response and policy intervention. Bangladesh J Bioeth 2(2):4–19
Islam MK, Kibriya NA, Dustegir MM (2018) Impact analysis of sand dredging from alluvial tidal river. In: E3S web of conferences, vol 40. EDP Sciences, p 03036. https://doi.org/10.1051/e3sconf/20184003036
Momen MN (2021) Regulatory governance and its significance in achieving sustainable development goals. In: Leal Filho W, Azul AM, Brandli L, Lange Salvia A, Özuyar PG, Wall T (eds) Peace, justice and strong institutions. Encyclopedia of the UN sustainable development goals. Springer, Cham. https://doi.org/10.1007/978-3-319-71066-2_103-1
Momen MN, Akhter MS, Begum MM (2024) Health sector performance in Bangladesh: challenges for sustainable development goals. In: Leal Filho W, Ng TF, Iyer-Raniga U, Ng A, Sharifi A (eds) SDGs in the Asia and Pacific region. Implementing the UN sustainable development goals—regional perspectives. Springer, Cham. https://doi.org/10.1007/978-3-030-91262-8_94-1
Mutton D, Haque CE (2004) Human vulnerability, dislocation, and resettlement: adaptation processes of river-bank erosion-induced displaces in Bangladesh. Disasters 28(1):41–62. https://doi.org/10.1111/j.0361-3666.2004.00242.x
Okeke CAU, Ede AN (2019) Mechanisms of riverbank failure and channel instability on the Nkisi River, Southeast Nigeria. IOP Conf Ser Mater Sci Eng 640:012104. https://doi.org/10.1088/1757-899x/640/1/012104
Paul SK, Routray JK (2013) An analysis of the causes of non-responses to cyclone warnings and the use of indigenous knowledge for cyclone forecasting in Bangladesh. In: Leal Filho W (ed) Climate change and disaster risk management. Climate change management. Springer, Berlin. https://doi.org/10.1007/978-3-642-31110-9_2
Piguet E (2013) From 'primitive migration' to 'climate refugees': the curious fate of the natural environment in migration studies. Ann Assoc Am Geogr 103(1):148–162. https://doi.org/10.1080/00045608.2012.696233

Rahman SA, Gain P (2020) Adaptation to river bank erosion induced displacement in Koyra Upazila of Bangladesh. Prog Disaster Sci 5(2020):100055. https://doi.org/10.1016/j.pdisas.2019.100055

Reid H, Huq S (2007) Community-based adaptation. A vital approach to the threat climate change poses to the poor. Available at https://www.iied.org/17005iied

Walsham M (2010) Assessing the evidence: environment, climate change and migration in Bangladesh. International Organization for Migration (IOM). Available at https://publications.iom.int/system/files/pdf/environment_climate_change_bangladesh.pdf

MIX
Papier aus verantwortungsvollen Quellen
Paper from responsible sources
FSC® C105338

If you have any concerns about our products,
you can contact us on
**ProductSafety@springernature.com**

In case Publisher is established outside the EU,
the EU authorized representative is:
**Springer Nature Customer Service Center GmbH**
**Europaplatz 3, 69115 Heidelberg, Germany**

Printed by Libri Plureos GmbH
in Hamburg, Germany